An Introduction to Rural Geography

An Introduction to Rural Geography

Andrew W. Gilg
Senior Lecturer in Geography
University of Exeter

Edward Arnold
A division of Hodder & Stoughton
LONDON NEW YORK MELBOURNE AUCKLAND

© 1985

First published in Great Britain 1985
Reprinted 1989, 1991

Distributed in the USA by Routledge, Chapman and Hall, Inc.
29 West 35th Street, New York, NY 10001

British Library Cataloguing in Publication Data

Gilg, Andrew W.
  An introduction to rural geography.
  1. Rural geography
  I. Title
  910′.091734   GF127

  ISBN 0-7131-6430-1

Typeset in 10/11 pt Times Compugraphic by Colset Private Ltd,
Singapore
Printed and bound in Great Britain for Edward Arnold, a division
of Hodder and Stoughton Limited, Mill Road, Dunton Green,
Sevenoaks, Kent TN13 2YA by Page Bros, Norwich

Contents

List of figures   vii
List of tables   ix
Preface and acknowledgements   xi
List of abbreviations   xiii

1 Introduction   1
Geography and rural geography — The scope and content of rural geography

2 Agricultural geography   9
Data sources — Regional change, classification and regionalization — Structure and employment — Decision making in agriculture — Agricultural geography and agricultural location theory — Concluding remarks

3 Forestry, mining and land use competition   38
Forestry — Mining — Land use competition

4 Rural settlement and housing   47
Theoretical approaches to rural settlement — Rural settlement policies — Rural housing

5 Rural population and employment   67
Rural population change — Rural communities — The classification and definition of rural areas — Employment in rural areas

6 Rural transport, service provision and deprivation   97
Rural public transport and accessibility — Rural service provision — Rural deprivation

7 Rural recreation and tourism   108
Recent trends — Impact studies — Second homes — Theoretical and quantitative approaches to rural recreation — Rural recreation geography: An emerging or mature discipline?

8 Land use and landscape   121
Land use studies — Landscape studies — Land and landscape evaluation

9   Rural planning and land management   138
Rural planning roles — Resource development roles — Conflict
resolution roles — Land management and integrated rural
development

10 Rural geography: Some concluding questions   171
Is there a discipline of rural geography? — What are the common
themes and questions in rural geography? — What direction might
rural geography take?

Bibliography   175

Index   200

List of figures

1.1 The 'inductive' and 'scientific' routes to explanation   3
2.1 Regional uptake of agricultural subsidies   14
2.2 Regional changes in type of farming in Scotland   15
2.3 Regional changes in type of farming Great Britain   17
2.4 Agricultural enterprise combinations in England and Wales 1958   18
2.5 Leading and second-ranking agricultural enterprises in enterprise
     combinations: England and Wales 1970   32/33
2.6 The Von Thünen model of agricultural location   35
3.1 Area of new forest by decade of planting   39
4.1 Development of central place theory and concept of thresholds   50
4.2 Theoretical rationalization of the settlement pattern   52
4.3 Key settlement plan for Devon   54
4.4 Settlement policies employed by County Structure Plans   55
4.5 Unit costs and costs per head of providing rural services   58
4.6 Rural housing profiles for England and Wales   62
5.1 Population change by counties in USA 1970–80   73
5.2 Percentage population change for Great Britain 1971–81   75
5.3 Absolute population change for England and Wales 1961–71 (left)
     and 1971–81 (right)   77
5.4 Two related models for urbanization and population change in the
     countryside   78
5.5 Cloke's index of rurality   86
5.6 Webber and Craig's classification of rural areas   88
5.7 Absolute and relative employment change 1971–77   94
6.1 The time/space realm of a rural housewife and various rural trans-
     port policy options   102
6.2 Two related models of the relationship between urban and rural
     deprivation   106
7.1 Key elements in the demand for outdoor recreation   110/111
7.2 Examples of origin–destination and recreation site surveys   113
7.3 Actual and theoretical relationship between site return rate and
     visitor frequency   119
8.1 Land use changes in Britain, Europe and North America   124/125
8.2 Land use change in the USA   126
8.3 Changing landscapes in the lowlands   129
8.4 Land use change in the uplands   131
8.5 Different purposes and units of data collection for landscape
     evaluation   135
8.6 A multivariate method of landscape evaluation   136
9.1 Goals of agricultural policy   141

viii   *List of figures*

9.2   Regional incentives and rural development agencies 1983   147
9.3   An example of a Structure Plan   153
9.4   Increase in built up area of the USA 1958–67   158
9.5   Ranked hierarchy of nature conservation sites and different
      management strategies   160
9.6   Protected landscapes and nature reserves in the United Kingdom   161
9.7   An example of a National Park Management Plan: The case of
      Dartmoor   162
9.8   Problem rural landscapes   165
9.9   Methods for improving the wildlife habitat and landscape of a low-
      land farm   166

List of tables

1.1	Recent publications in rural geography	2
1.2	Categories of rural research	5
1.3	Subjects of rural geography research 1973–81	6
1.4	Research interests of Commonwealth rural geographers	7
1.5	Contents of two recent books on rural geography	8
2.1	Type of information provided by the agricultural census	10
2.2	Regional farm-type specialization	13
2.3	Weaver's method of producing crop-combination regions	16
2.4	Distribution of holdings and standard man-days by type of farming	21
2.5	Structure and distribution of holdings by size and enterprise 1975	22
2.6	Area and numbers of holdings rented or owned 1960–75	23
2.7	Change in the number of enterprises in central Somerset 1963–76	24
2.8	Change in the type of enterprise in central Somerset 1963–76	24
2.9	Attractions of farming life suggested by East Anglian farmers	28
2.10	Goals and values of East Anglian farmers and West Midlands hop farmers	28
2.11	Ranking of agricultural decision making factors by order of importance	29
2.12	Newby's typology of East Anglian farmers	30
3.1	Conclusions of the 1972 cost/benefit study of forestry	41
3.2	Production of minerals in the UK 1950–70	43
4.1	Change in village size 1563–1911	48
4.2	Characteristics of central places in southern Germany	49
4.3	Savings to be obtained from relocation depending on rate of discount	51
4.4	Population change in the Bury St Edmunds area 1961–76	57
4.5	Some indices of rural housing	60
4.6	Pahl's typology of social groups relevant to rural housing	61
4.7	The character of the rural housing profiles shown in Figure 4.6.	63
4.8	Housebuilding in the countryside	65
5.1	Summary of factors likely to cause migration	68
5.2	Population change by size and type of settlement	74
5.3	Population change by type of district	76
5.4	Summary of rural population change	79
5.5	Socioeconomic characteristics of life in a metropolitan village	82
5.6	Selected characteristics of urban and rural areas in England and Wales	85
5.7	Reasons for mobility from Norfolk farms 1960–70	90
5.8	Trends in employment sector for selected counties	91
5.9	Youth employment, aspiration and experience in mid-Devon	92

5.10 Different costs of providing various rural employment options   92
6.1   Some features of rural transport   98
6.2   Principal transport policy changes and their impact on rural areas 99
6.3   Overall definition of public transport needs in rural areas   101
6.4   Number of rural settlements by size and distance from an urban
      centre   103
6.5   Access to a car by social/age groups   103
6.6   Inadequate levels of rural service provision   104
7.1   Changes in factors and trends in rural recreation   109
7.2   Visitors' attitudes to two types of recreational site   114
7.3   Concentration of recreation   115
7.4   Second homes in Europe and North America   116
7.5   Costs and benefits of second homes   116
8.1   Land use within Britain: A summary of the principal data sources 122
8.2   Land use structures of England and Wales, Europe and USA   123
8.3   Changes in lowland agricultural landscapes 1945-72-83   128
8.4   Enduring conversion of rough pasture   130
8.5   Changes in upland landscapes 1967-78   133
9.1   Aims for rural planning 1944 to 1977   139
9.2   Pattern of public expenditure on agriculture   143
9.3   EEC Intervention stocks and consumption   143
9.4   Key agricultural data for EEC and USA   144
9.5   World trade in farm products   145
9.6   Types of agricultural policy measures   146
9.7   Expenditure by development agencies   149
9.8   Planning applications and protected landscapes   157
9.9   Past and possible future costs of management agreements in SSSIs 168

Preface and Acknowledgements

When I was asked to write this book, I was at first hesitant, not only because it would be my third book on countryside issues in six years, and because their writing always seemed to coincide with either very hot long summers (1976 and 1984) or major cricket tours by the West Indies (1976 and 1984) or Australians (1978), but mainly because, after a gap of over 10 years, a number of other books on rural geography were due to appear. My hesitation has in fact been a blessing in disguise, since these books had all appeared by the time I began to write and I have hopefully been able to provide a complementary coverage, so that the whole field of rural geography is adequately covered by the books now available.

For example, Pacione's *Rural Geography* (1984) provides a systematic account of the geography of rural areas, Phillips and Williams provide a sound introduction to the social geography of 'Rural Britain', and the individual contributors to 'Progress in Rural Geography' provide reviews of some of the main sub-disciplines within rural geography. What is missing therefore is an overview of not only the content of rural geography, but the approach of rural geographers to their subject, particularly with regard to data collection, methods of description and explanation, the development of theory, and the application of their results to policy-making.

Accordingly, this book is intended to provide a review of what rural geographers do, how they do it, what they have found out, and how they have related their findings to the real world of policy formulation and decision making. This means that the book is focused around a review of over 750 references and should provide a most comprehensive reference source. Many of the references are UK derived, since this reflects the literature available. However there are also, where relevant, extensive references to Europe and the USA, when either the references are more generally available there or when the subject area differs markedly from experience in the UK. Nonetheless, there are topics where the literature, in spite of the information explosion about rural areas in recent years, is less than adequate, and when this is the case a comment is made in the text.

In writing the book I have therefore been indebted to the authors of all these references, many of whom are professional colleagues and friends, and I do hope I have done their work justice. I am also very grateful to my 'rural' colleagues at Exeter, Mark Blacksell, Mark Cleary, David Phillips, Gareth Shaw and Allan Williams for advice and encouragement, and in particular to Andrew Teed, Rodney Fry and Terry Bacon for all their technical help in preparing the diagrams and maps for publication. Finally, I must thank my wife, Joyce, and two children, Julie and Alastair for giving up much of a long hot summer holiday while the book was written. Their compensation has been winter holidays in the Alps, where the multiple land use message and applied rural geography theme of this book has been put most effectively into practice, most notably at Crans Montana, which with Murrayfield, inspired the author through the times of doubt that all authorship involves.

In conclusion, the following publishers, journals and organizations are acknowledged for allowing the following figures to be reproduced in this book: Allen & Unwin and Allan Rogers (4.6); Edward Arnold (4.1 and 8.2); Countryside Commission 8.3, 8.4, and 9.9); Croom Helm (7.1 and 8.1); David & Charles (2.6, 4.2, 4.3, 8.5 and 8.6); Geografiska Annaler (5.4); Geographical Review (9.4); Faber & Faber (2.4, 2.5); Journal of Agricultural Economics and Ian Bowler (2.2); Longman (7.1); Methuen (2.3, 6.1, and 8.1); Progress in Planning (4.5); Regional Studies Association (5.5); Scottish Geographical Magazine (5.1); Town and Country Planning (5.2); and Town Planning Review (4.4).

Exeter
September 1984

Abbreviations

AAAG	Annals of the Association of American Geographers
ACC	Association of County Councils
ADAS	Agricultural Development and Advisory Service
ADC	Association of District Councils
BTA	British Tourist Authority
CAP	Common Agricultural Policy
CAS	Centre for Agricultural Strategy
CCS	Countryside Commission for Scotland
CCP	Countryside Commission Publication
CLA	Country Landowners Association
COSIRA	Council for Small Industries in Rural Areas
CRC	Countryside Review Committee
DAFS	Department of Agriculture for Scotland
DART	Dartington Amenity Research Trust
DBRW	Development Board for Rural Wales
DES	Department of Education and Science
DOE	Department of the Environment
EC	European Community
EEC	European Economic Community
FWAG	Farming and Wildlife Advisory Group
GNP	Gross National Product
HC	House of Commons
HIDB	Highlands and Islands Development Board
HL	House of Lords
IBG	Institute of British Geographers
ITE	Institute of Terrestrial Ecology
IUCN	International Union for Nature Conservation
MAFF	Ministry of Agriculture, Fisheries and Food
MOH	Ministry of Health
NCC	Nature Conservancy Council
NCVO	National Council for Voluntary Organizations
NFU	National Farmers' Union
NPA	National Park Authority
OECD	Organization for Economic Cooperation and Development
OPCS	Office of Population Censuses and Surveys
RASE	Royal Agricultural Society of England
RCC(s)	Rural Community Council(s)
RICS	Royal Institute of Chartered Surveyors
SDD	Scottish Development Department
STB	Scottish Tourist Board
TRRU	Tourism and Recreation Research Unit
WWF	World Wildlife Fund

To Jim Aitken and his Scottish Rugby Union team winners of the
Grand Slam 1984

21-1-84	Cardiff	Wales	9	Scotland	15
4-2-84	Murrayfield*	Scotland	18	England	6
3-3-84	Dublin	Ireland	9	Scotland	32
17-3-84	Murrayfield	Scotland	21	France	12

* Edinburgh

1

Introduction

Geography and rural geography

The desire to know and the compulsion to tell are the *raisons d'être* of any academic subject. Exactly what it is desired to know and how it should be described, analysed and explained provide the artificial limits that demarcate one academic discipline from another. In this sense all subject areas are capable of having their boundaries redefined, and this is why, when considering an area of knowledge as wide and diverse as rural geography, it is necessary to consider at the outset not only the changing nature of rural geography but also the changing nature of its host discipline, geography.

Rural geography as measured by the rate of publication (with three new books and two new journals in the early 1980s as shown in Table 1.1) has experienced a rebirth within geography since 1970 and has once again asserted itself as a major sub-discipline within the overall subject area. This is in fact a partial return to an earlier period when virtually all geography was rural geography since the world was dominantly rural. Until the 1850s, and even later in some areas, geography was dominated by need to explore the world's unknown regions. Furthermore, the subject could be conceived of as a whole and any one scholar could encompass all the written geography of the period. As more information became available the subject was forced to divide into sub-disciplines and to develop theories to classify and explain the increasing weight of available data (Johnston, 1979).

Following on from the work of Darwin many geographers proposed the theory that man's actions were determined by the physical environment. Obviously this theory was most attractive in rural areas where farming systems are by their very nature closely related to the physical landscape. These theories of environmental determinism were then followed up by the theory of natural regions and the inter-war years were dominated by the regional concept. In rural areas Vidal de la Blache used the French word '*Pays*' to describe distinctive tracts of countryside which could be physically delimited by their appearance, and Hartshorne developed this idea further to state that geography was concerned with the 'areal differentiation' of the earth's surface. In this period geography was firmly rooted in the regional landscape school and the only real differences were about the methods employed. For example, Sauer believed in studying landscape as an ongoing generic process of change, while Wooldridge and East argued that geographers should synthesize knowledge about a region into a coherent set of knowledge. But all these approaches of necessity placed the rural landscape and its study at the centre of geography.

In the 1950s and 1960s, however, the regional paradigm was gradually replaced by systematic studies which didn't cover any specific geographical

Table 1.1: Recent publications in rural geography

Type of publication	Date	Title	Author/editor
Annual journal	1980–1984	*Countryside Planning Yearbook* Volumes 1–5	(Gilg, 1980–1984)
Edited collection of papers	1983	*Progress in Rural Geography*	(Pacione, 1983)
Textbook	1984	*Rural Britain: A Social Geography*	(Phillips and Williams, 1984)
Textbook	1984	*Rural Geography*	(Pacione, 1984)
Quarterly journal	1985	*Journal of Rural Studies*	(Cloke, 1985)

area, and were conducted at all scales from the macro to the micro. In addition, in a classic paper, Schaefer (1953) wrote that claims to be able to synthesize all the available knowledge in any one area were arrogant. He argued instead that the natural sciences were characterized by the pursuit of explanation, and that explanations required laws. Schaefer's arguments were the natural forerunner for the method of enquiry now known as either the scientific method or logical positivism as shown in Figure 1.1. This is based on the concept that there is an objective world in which there is order waiting to be discovered. The basic method of logical positivism presupposes the development of a theory and then its testing by empirical data collection. This method rapidly gained ground in the 1960s and aided by the development of computers, statistical methods and explanatory models it marked out the 1960s as the period of the so-called quantitative revolution in geography. In contrast to the earlier periods of environmental determinism and regional geography this 'new geography' was most relevant to economic and urban geography, and it is largely for this reason that rural geography, the mainstay of human geography till the 1950s, lost its place in the mainstream of geography.

There were however some areas where the onset of the logical positivist approach was directly relevant to rural geography. In particular the development of 'location theory' and the concept of the 'friction of distance' allowed rural geography to improve its contribution to the study of rural settlement, rural transport and rural recreation and this contributed to a shift away from the study of rural landscape to the study of rural economy and society. Another facet of the new geography, the 'systems approach', was however less relevant to rural geography and increasing dissatisfaction with the rather restricted nature of the logical positivist approach encouraged a variety of new approaches to geography in the 1970s and early 1980s, and in particular, behavioural geography, cultural geography, humanistic geography and radical geography.

All of these, to a greater or less extent, were more applicable to rural geography than the scientific methods of the 1960s and this helps to explain the remarkable resurgence of the subject from 1970 onwards. For example, behavioural geography, which recognizes that man's actions are not always rational, allowed new studies of farmers' decision making processes, and cultural geography recognized the basic fact that the world isn't an isotropic plain and that regional differences and unique rural landscapes and rural

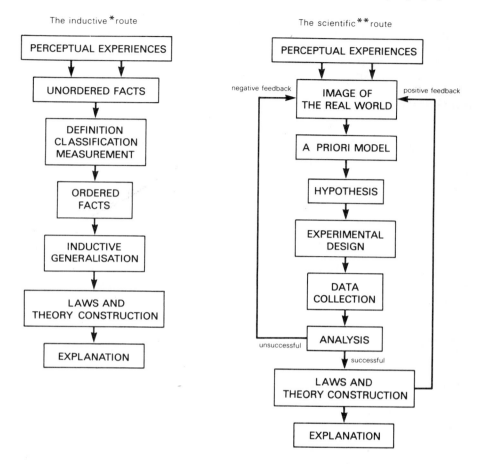

Fig 1.1 The 'inductive'* and 'scientific'** routes to explanation. Sometimes known as the 'empirical'* and 'logical positivist'** routes respectively. Adapted from Harvey (1969)

societies are not only very important, but intrinsically interesting in their own right. Both these approaches can be bracketed under the humanistic geography approach which seeks to study how people behave individually rather than by comparing normative models with observed data.

Another change in this area has been the development of radical geography, in which geographers have moved away from their traditional apolitical stance and (arguing from either liberal or marxist standpoints) have sought to study not how the world is, but how it should and could be. This approach is reflected in the field of rural geography by studies of rural deprivation, and by the career of one very eminent geographer David Harvey. His first work, an unpublished PhD thesis, dealt with *Aspects of agricultural and rural change in Kent*

1800-1900 and argued that physical factors were only one explanatory variable in agricultural geography and that socioeconomic factors were just as important (Harvey, 1963). In spite of some theoretical review work on the analysis of agricultural land use patterns (Harvey, 1966), Harvey, however, soon left rural work and in two further shifts argued first for the 'logical positivist' approach in *'Explanation in Geography'* (Harvey, 1969) and then in a complete *volte face* for radical geography in *'Social Justice and the City'* (Harvey 1973), thus completing a remarkable shift in approach in a mere decade.

The resurgence of rural geography, although partly due to the development of humanistic geography, and the spread of the logical positivist approach to rural studies, has been due perhaps even more to the rise of environmentalism and a growing concern about man's effect on the environment, a sort of environmental determinism in reverse. Although many environmental issues are urban-based, for example, air pollution, the majority of such issues are rurally based and rural geography has taken full advantage of this new interest in the environment. Accordingly, environmentally based rural geographers have become interested and involved in studies of landscape change and conservation, recreational land use and management, and managing and planning the rural resource base. In these areas of interest, rural geographers have employed all types of approach, but the most common and widely accepted is still the logical positivist approach, as the next section and the succeeding chapters will show.

The scope and content of rural geography

It is not the intention of this book to become bogged down in internal and infernal discussions about the definition of rural areas although this topic is covered in more detail in Chapter 5. Nonetheless a definition is needed if only for purely pragmatic reasons. Many of the definitions in the literature rest on ideas of population size, distance from a large town – for example, a radius of five miles from a town of 20,000 people and 10 miles from a town of 100,000 (Norfolk County Planning Dept., 1979) – or the degree of urban influence, but the most satisfying and complete definitions are those based on the appearance of the landscape and the intensity of land use. In this respect the most satisfactory definition remains the one proposed by Wibberley (1972, 259) who argued that the word rural:

> describes those parts of a country which show unmistakeable signs of being dominated by extensive uses of land either at the present time or in the immediate past

This definition has been widely used by most rural geographers and also accepted by other rural workers, for example, rural sociologists (Miller and Luloff, 1981, 609-10) who note that:

> the term 'rural' is conventionally employed to denote a delimited geographical area characterized by a population that is small, unconcentrated and relatively isolated from the influence of large metropolitan centres

Following this approach the first rural geography text to emerge after the lacuna of the 1960s, Clout's *'Rural Geography'* (1972, 1) defined rural geography as:

> The study of recent social, economic, land use, and spatial changes that have taken place in less-densely populated areas which are commonly recognized by virtue of their visual components as countryside

The use of the word 'recent' effectively removes the historical study of rural settlements, one of the mainstays of former rural geographies, but retains the study of agriculture which had also dominated the subject until the 1960s. Indeed, until 1974 the Rural Geography Study Group of the Institute of British Geographers (IBG) had been the Agricultural Geography Study Group (Clark, 1982a). The widening, if ununified approach to rural geography throughout the 1960s, was recognized in Clout's 1972 book in which he listed about 20 components of rural geography ranging from rural employment to water gathering. In spite of the fact that Clout grouped these components into three key themes, outmigration from the countryside, repopulation of some rural areas, and planned and unplanned change in the countryside, he recognized that this was an eclectic choice rather than any attempt to build up a comprehensive framework for rural geography. Indeed, even five years later Clout (1977, 475) was able to observe of the early rural literature that:

> no specific methodology binds them together and they are united only by the fact that they are concerned with the less densely occupied sections of the earth's surface

In spite of attempts to develop epistemologies, ideologies, paradigms and other associated -ologies, rural geography remains in common with other branches of the subject obstinately eclectic and theory-free (Clark, 1982a) and above all rural geography is still what rural geographers do. Accordingly, the most effective way to discuss the scope and content of rural geography is to review the various research registers of rural geography.

The first register to be compiled by the new Rural Geography Study Group of the IBG (Bowler, 1975) classified the work of rural geographers according to the categories shown in Table 1.2. This shows not only the wide range of the subject but the dominance that agricultural studies still had in the early 1970s. However, subsequent registers in 1978 and 1981 (Clark, 1978; 1981) revealed

Table 1.2: Categories of rural research

Agriculture	*Rural population*
— enterprises and farm systems	— distribution, growth, depopulation
— structural change	— employment structure
— land use patterns	— social geography of rural communities
— marketing and distribution	
— social geography of agriculture	*Rural transport*
— theoretical approaches	— model studies
	— impact studies
Forestry	
— recreational use	*Recreation and tourism*
— social and employment impact	— second homes
— economic aspects	— resorts
— conservation	— population movement
	— land use effects
Rural settlement	— national and country parks
— economic structure or function	— demand assessment studies
— social structure of settlements	
— planning of settlements	*Rural (development) planning*
— historical development of settlements	— industrial development
	— rural resource evaluation
	— institutions
	— planning control of land use

Source: Bowler (1975).

Table 1.3: Subjects of rural geography research 1973–81

Percentages	1973–4	1978	1981
Agriculture	40.6	40.2	27.5
Forestry, fishing, conservation and mineral extraction	1.6	4.0	8.7
Rural settlement	19.7	16.1	16.5
Rural population	12.0	10.3	15.1
Rural transport	3.6	4.5	4.6
Recreation and tourism	6.8	11.6	8.3
Rural development and planning	15.7	13.4	19.3

Source: Clark (1982a).

significant changes in the content and scope of rural geography as the 1970s progressed, as shown in Table 1.3. The relative decline in the importance of agricultural studies has been matched by more research in the previously neglected fields of fishing and forestry and by a new interest in conservation (Clark, 1982a). The other main changes have been a growing interest in social and economic affairs and rural development and planning. These developments are partly mirrored by the career of the eminent rural geographer, Terry Coppock. In common with David Harvey, Terry Coppock's first papers dealt with land use change (Coppock, 1960a), he then developed this work to produce one of the first computer-derived atlases: 'The Agricultural Atlas of England and Wales' (Coppock, 1964a). But since then, Coppock's career has branched out in common with the flowering of rural geography to include recreation (Coppock and Duffied, 1975), resource management (Coppock and Sewell, 1975) and conservation (Coppock, 1970).

The registers produced by the Rural Geography Study Group, although not confined to the British Isles, are dominated by studies of the United Kingdom, with 60 per cent of the 1981 studies being conducted there. A wider perspective is provided by a 1982 Review and Directory of Rural Geography in the British Commonwealth (Wood and Smith, 1982). As Table 1.4 shows, the pattern of research interest is similar to the British survey and confirms that even in the Third World rural geography is more than the economic geography of agriculture. Although agriculture continues to dominate, there are nearly as many studies collectively of rural settlement and rural development and planning in the developing world (23 and 33) as there are of agriculture (59). Both sets of registers point out that research is becoming increasingly applied and that empirical work based on the case-study approach is the dominant mode of operation. In contrast there has been little development of theoretical frameworks and quantitative methods in comparison to other branches of geography, notably urban and economic geography.

These conclusions are confirmed by the 10 authors of a 1983 text, *Progress in Rural Geography* (Pacione, 1983), and in his introduction Pacione (1983, Introduction) writes:

> Rural geography has a long tradition but until recently it generally referred to studies concerned with agriculture or comprised historical analyses and descriptions of the settlement or land-use patterns of the countryside. Although these areas of investigation retain their importance within the subject today, rural geography has expanded over the last decade to encompass other lines of enquiry such as the systematic study of rural transportation, employment and housing; assessments of

Table 1.4: Research interests of Commonwealth rural geographers

	Percentage distribution among all topics	Distribution of topics between industrialized and developing world
Agriculture	38.0	48:59
Non-agricultural activities	1.8	2:2
Recreation and tourism	1.8	6:0
Rural environment	5.6	4:10
Rural population	8.2	8:12
Rural settlement	18.0	33:23
Rural infrastructure	3.0	2:4
Health and welfare	2.2	2:3
Rural development and planning	21.4	28:33

Source: Wood and Smith (1982).

development policies in rural areas; and attempts to develop theory and methodology in rural studies

It is indeed the last statement that casts the only real shadow on the remarkable resurgence of rural geography since 1970, for as Cloke (1980) has argued in an excellent review of the subject, rural geography has reached an impasse with regard to its theoretical and methodological development. He therefore urges that among the new emphases for rural geography the first task should be, in general, the development of a conceptual framework, and in particular, the answer to two fundamental questions:

1   What is 'rural' and how is it different from 'urban'? and
2   What spatial dynamics are involved in the manifestation of this difference?

Cloke also suggests other new emphases, including the isolation of suitable analytical tools, since present-day rural geography is hampered not only by conceptual famine but also by methodological plagiarism and misuse of statistical techniques. The final new emphasis should be an integrated approach for rural geography and planning, because the division of rural geography into the neat subsections of transport, agriculture, conservation, etc. has paid insufficient regard to their interconnexivity.

The weak points of rural geography outlined by Cloke are not unique, however, to rural geography, or even other branches of geography, and are also reflected in other disciplines. For example, in rural sociology, a related discipline, a literature survey of the main journal *Rural Sociology* from 1965 to 1976 revealed just the same lack of a conceptual framework, and produced much the same solutions advocated by Cloke, namely a new rural sociology based on theoretical and methodological pluralism, in terms of the changing structure of rural societies and the many dimensions of the concept 'rural' (Picou *et al.*,1978).

In conclusion, this chapter has outlined how rural geography has reasserted itself as a major sub-discipline within geography, but has also shown that it has re-emerged as a different creature with new concerns and interests as shown in Table 1.5. It is however still based heavily in the empirical and case-study approach and although it has exponents among many of the new geographical sub-paradigms, for example, behavioural and radical geography, it remains theoretically and methodologically undeveloped.

Table 1.5: Contents of two recent books on rural geography

Pacione, 1984	Phillips and Williams, 1984
1 Introduction	1 Rural social geography
2 Evolution of the settlement pattern	2 The rural economy i: Living off the land
3 Spatial organization of settlement	3 The rural economy ii: Non-agricultural
4 Settlement planning and change	employment in rural areas
5 Agriculture in the modern world	4 Population and social change
6 Structural change in agriculture	5 Housing
7 Agriculture and urban development	6 Transport and accessibility
8 Population dynamics	7 Planning within the countryside
9 Rural communities	8 Services and retailing
10 Metropolitan villages	9 Recreation and leisure
11 Seasonal suburbanization	10 Deprivation
12 Quality of life	11 Policy issues and the future
13 Housing	
14 Employment and rural development	
15 Service provision	
16 Transport and accessibility	
17 Resource exploitation and management	
18 Conservation	
19 Leisure and recreation	
20 Power and decision making	

Sources: Pacione (1984) and Phillips and Williams (1984).

2

Agricultural geography

Even today agriculture remains a major and the most areally extensive industry, not only throughout the less developed world but also in heavily urbanized areas like Western Europe, where it still accounts for 2.2 per cent of Britain's Gross Domestic Product (MAFF, 1983a). Clearly a comprehensive study of all aspects of agriculture would fill many volumes and so this chapter concentrates on those parts of agriculture where rural geographers, and allied workers in other subject areas like agricultural economics have spent most research time, or contributed the most effective data or theory. Although this is a matter of opinion, their work can be divided into *data sources, classification and regionalization, structure and employment, decision making* and *location theory*, and in fact this division fairly closely accords with Bowler's (1984) grouping of traditional research in agricultural geography into data sources and regionalization, farming systems, agricultural resources and decision making. In spite of the fact that some of the research work reviewed here could be put into another section, this chapter covers each of the aspects italicized above in turn, and then concludes by considering if there is a coherent discipline of 'Agricultural Geography'.

Data Sources

Because agriculture by its very nature is an extensive land use, it is impossible for any one worker, or even team of workers, to collect land use or other farm data except for over a limited period or area. Fortunately most governments conduct an annual or frequent census of agriculture, indeed these are usually far more frequent than censuses of population or industrial production.

In Britain there have been annual censuses of agriculture since 1866 (MAFF, 1968) and this has allowed geographers to plot long-term trends with relative ease. The data are available from the Ministry of Agriculture at parish and Agricultural Development and Advisory Service district level (about 30 to 40 parishes combined) and are also published by HMSO at the county level (MAFF, 1983b). The data are extraordinarily comprehensive but deal solely with the exact areas of crops, of land rented or owned, the numbers of farmers and workers, and the total number of livestock as shown in Table 2.1. The data do not provide any information about the processes of agriculture change or why certain crops or livestock are produced. Accordingly, the data are very useful as a descriptive device but are relatively useless as an explanatory tool.

Some analysis is however provided by another government source, the *Annual Review of Agriculture* (MAFF, 1983a). This 'Review' provides a brief summary of trends in agriculture and an aggregated set of data on the economic

Table 2.1: Type of information provided by the Agricultural Census in the UK

	1979	1982	% Change
Land use and tenure			
('000 hectares)			
Wheat	1,372	1,664	+ 21
Barley	2,347	2,221	− 5
Potatoes	204	191	− 11
Sugar beet	214	204	− 5
Total crops	4,914	5,065	+ 3
Total crops and grass	12,100	12,071	0
Total rough grazing	6,352	6,210	+ 2
Woodland on farms	264	285	+ 8
Total area	17,725	17,568	− 1
Land rented	7,028	6,730	− 4
Land owned	10,697	10,838	+ 1
Livestock (thousands)			
Cattle (breeding herd)	4,835	4,647	− 4
Total cattle	13,589	13,275	− 2
Pigs	7,864	8,082	+ 3
Sheep and lambs	29,946	33,049	+ 10
Labour (thousands)			
Whole-time farmers	216	203	− 6
Total farmers	304	295	− 3
Regular whole-time workers	187	170	− 10
Regular part-time workers	66	62	− 6

Source: Gilg (1981 and 1983a).

and production state of the industry. Since 1973 Britain has been a member of the European Economic Community and so European statistics are also a useful data source (EC, 1980), as on the world scale are the various annual publications on world agriculture compiled by the United Nations or by its constituent Food and Agricultural Organization.

Agricultural data at the world or local level are however only as good as the database from which they are aggregated, and in Britain the database has a number of built-in inaccuracies – for example, farmers may simply copy out the previous year's returns (Clark *et al.*, 1983). A more serious limitation is the spatial unit used to collect the data, the parish, and according to Coppock (1960b) the most serious drawback is the fact that farm holdings do not share the same boundaries as parishes. This means that some parishes overstate their agricultural area while others underrecord their farm area. Other problems relate to the physical heterogeneity of parishes, their great range of size and shape, and in particular their lack of relationship with any of the other factors in agricultural geography – for example, they often cut clean across significant boundaries between different soil types. Coppock therefore concluded that the parish was an unsatisfactory unit for the study of agricultural geography, but since there was no alternative, that the only satisfactory procedure was to take steps to minimize its disadvantages, by for example judicious amalgamation.

In view of the disadvantages of publicly collected data a number of studies have either adapted them, used other data or collected their own data, although Coppock (1960a) in his study of crop and livestock changes in the Chilterns between 1931 and 1951 placed great reliance on the unpublished parish summaries of the agricultural returns, since there were few other documentary sources. In a more recent study of land use change in the southern United States Hart (1980) noted that US Census of Agriculture returns showed a staggering fall in farmland from 6,340,000 hectares in 1939 to 2,495,000 hectares in 1974. In order to explain, albeit controversially (Trimble, 1983), this amazing change Hart had to resort to a number of other data sources, namely published maps of land use, county tax records, case studies of individual farmers, and a comparison of air photos taken in 1937 and 1974. This allowed Hart to conclude that the main reason for the land use change was the reversion of cotton land to scrubby woodland. His air photo data for one county in Georgia revealed that 36,760 hectares had changed from agriculture to forest, while in contrast the US Census data revealed a loss of 23,150 hectares of forest during a similar period. This example clearly illustrates how official data can distort the real world situation as revealed by air photo data, and in this case reflects both the farmers' view that scrubby woodland did not represent farmed woodland and was therefore left off from their returns to the agricultural census, and the fact that a good deal of such land had been sold off to non-agricultural owners, and thus once again did not appear in the agricultural census.

In the light of Hart's results it is not surprising that geographers have turned to other data sources and in particular to vertical air photographs and related remote sensing techniques, notably satellite imagery.* However, remote sensing suffers from a number of drawbacks. The most important is the problem of interpretation, for not only is the field of view unfamiliar but the patterns, especially if infra-red or radar techniques are used, are very difficult to interpret for the full range of crops or as the seasons vary. Satellite imagery also suffers from the problem that it is only accurate enough for mapping at scales above 1:50,000 (Royal Geographical Society, 1983). Other difficulties relate to the cost of the exercise, the static one-off nature of the record presented, cloud cover, and for satellite imagery political sensitivity (Allan, 1980). So far therefore all remote sensing techniques have had to rely on extensive ground-based land use surveys to provide an *in situ* control and this has added further to the expense and complexity of the approach (Barrett and Curtis, 1976).

Accordingly land use surveys whether on a sample basis (Clark and Gordon, 1980) or a complete coverage still have an important role to play in providing data for agricultural and rural geographers. There have been three main land use surveys of the British countryside. In the 1930s Sir Dudley Stamp organized the first national land utilization survey. Although the published maps at the 1:63,360 scale can only provide limited spatial detail and the 30 or so categories of land use shown are very broad, the accompanying County reports provide a wealth of material for anyone wishing to study the rural geography of the 1930s (Stamp, 1962).

The second land utilization survey organized by Alice Coleman on a complete coverage in the 1960s, and on a sample coverage in the 1970s and 1980s, has the advantages of both a larger scale of publication 1:10,560 or 1:10,000 and with over 60 land uses it also shows over twice as many land use categories (Coleman *et al.*, 1974). A problem with all land use surveys however is that they are purely descriptive, in common with agricultural census data. Nonetheless they can be analysed to provide more useful information. For example: agricultural data

---

* Remote sensing is also discussed in Chapter 8.

can be classified into significant groupings; regional differences can be mapped; the structure and employment patterns of agriculture can be described; and theories tested. Accordingly, the next two sections discuss the analysis of agricultural census data before the succeeding section attempts a discussion of decision making in agriculture and thus an explanation of the patterns revealed.

Regional change, classification and regionalization

One of the most traditional methodological approaches in agricultural geography has been the regional analysis of different farm enterprises, and their changing distributions through both time and space. Pocock's (1960) article on the changing regional patterns of hop growing, based on parish census data between 1866 and 1957 is typical of the *genre*, and describes a process of concentration into the two key areas of southeast England and the West Midlands.

Twenty years later studies of this kind are still being conducted but within a far more theoretical framework. For example, Ilbery's study of the decline of hop growing in Hereford and Worcester (Ilbery, 1982; 1983a) attempted to see if Hägerstrand's theory of the contagious adoption of an innovation (Hägerstrand, 1967) worked in reverse as proposed by Barker (1977) and to see if Harvey's socioeconomic principles (see page 26) (Harvey, 1963) still applied. Using data supplied by the Hop Marketing Board, Ilbery was able to plot the pattern of decline, relate this to geology, soil type and land use capability via the Chi-squared test, and conduct a questionnaire study of those farmers who had grubbed up their hops. From these data, Ilbery was able to conclude that there wasn't any evidence of a contagious process with neighbours grubbing up hops in turn, but that the smallest hop yards had been grubbed up first and that the main explanatory variables were social and economic, namely the uncertain future of the crop and its high capital cost. However, this had also led to a greater emphasis on the more suitable physical areas and so Harvey's economic principles were found to be of limited value (Ilbery, 1983a).

In spite of the decline of one crop, hops, the more common process in the last 30 years has been the adoption of new crop systems and farm enterprises, and in particular the role played by government grants and subsidies since as Chapter 9 demonstrates, agriculture is now closely directed by government policies. Bowler (1976a) has studied the take-up of a number of such subsidies, for example, the calf subsidy 1949–74, and the ploughing-up grant 1952–66. Although his work has revealed that these measures were subservient to other economic and technical trends, and only exerted short-term effects, another study by Bowler (1976b) has revealed a greater impact for regional policies in upland areas, but as Figure 2.1 reveals the impact was also spatially uneven. Bowler also found that production subsidies for hill cows and hill sheep had helped to concentrate these production systems in the eligible areas, and that investment grants for farm improvements had gone to the farms least in need of aid, and that the poorer farms had remained poor.

The most widespread regional policy in agriculture is however the regional pricing policy of the Milk Marketing Board (Baker, 1974), and this has distorted the regional pattern of production away from the centres of demand, to the more physically suitable areas, mainly by the modification of regional transport

Table 2.2: Regional farm-type specialization

Type of farm (per cent within each region)	Region 1975 Highland	North East	East Central	South East	South West	Scotland 1968	1975	Coefficient of localization C or D *
Hill sheep	10.9	0.2	3.6	4.7	3.7	5.2	4.2	D
Upland	21.6	10.1	12.0	10.2	27.0	15.8	24.7	D
Rearing with arable	5.9	22.3	8.5	12.4	4.4	12.3	14.1	D
Rearing with intensive livestock	0.3	2.6	0.9	0.9	0.6	2.1	1.5	C
Arable rearing and feeding	1.6	4.9	4.4	3.8	0.6	4.9	3.6	D
Cropping	2.8	6.9	31.6	19.3	1.8	13.9	11.7	C
Dairy	5.7	4.3	6.6	7.3	34.7	20.0	16.2	C
Intensive	0.8	2.1	8.3	6.4	4.5	4.7	4.7	D
Part time	42.9	17.2	8.5	8.9	9.8	21.2	19.3	C

*C = more concentrated.   D = more dispersed.

Coefficient of localization for the ith type of farm $= \sum\limits_j \dfrac{\left| \dfrac{e_{ij}}{\sum_j e_{ij}} - \dfrac{a_j}{\sum_j a_j} \right|}{2}$

where e_{ij} = number of farms of the ith type in the jth region
a_j = area of crops and grass in the jth region
0 = even distribution.   1 = concentrated distribution.
Source: Bowler (1981).

charges (Barnes, 1958; Simpson, 1959). However it is not wholly true to say that milk production is undertaken in those areas to which it is best suited, since the need to produce an adequate and regular income from a small farm has encouraged milk production on the upland fringe, where physical conditions are often far from ideal and real costs of transport are high (Coppock, 1971). Other criticisms of the regional pricing policy have been made on the grounds that it is against the public interest and unfair to producers in the southeast. But Rayner (1977) using a partial welfare economics framework and quantitative techniques to estimate the cross-subsidization implicit in the scheme, has concluded that the policy has not worked unduly against the public interest.

So far all the studies of change that have been examined have looked at one crop or enterprise or at most a limited range of agricultural systems, but of course geographers have also been interested in the changing pattern and interplay of different farming types. Bowler (1981) for example, has used farm classification data, produced by the Department of Agriculture for Scotland for 1968 and 1975, to show regional variations in farm type specialization as shown in Table 2.2 and Figure 2.2. If only the farm types accounting for 10 per cent or more of the farm total in 1975 are taken, it can be seen from Table 2.2 that the two types that increased their share – 'upland' and 'rearing with arable' – had become more dispersed, while those that had decreased their share – 'cropping' and 'dairy' – had become more concentrated. However, as Figure 2.2 shows the spatial pattern is more complex, and although Bowler concludes that for many enterprises and farm types, a greater regional localization is accompanying the

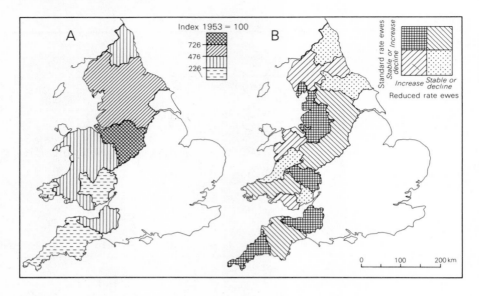

Fig 2.1 Regional uptake of agricultural subsidies. A: Subsidized hill cows 1953–73. B: Subsidized breeding ewes 1943–1966. Source: Bowler (1976)

trend to enterprise concentration, the process is by no means uniform.

Bowler (1983a) has also used shift and share analysis to show significant adjustments in the regional distribution of various types of farm between 1965 and 1973, as shown in Figure 2.3, in which positive values indicate absolute increases in farm numbers or rates of decrease below the national average, while negative values show rates of loss in excess of the national average. Bowler concludes that the evidence in broad terms shows a strengthening of traditional farming in core type regions, and a reduction of farm types atypical for a region. However, although the evidence suggests a trend towards a further regional localization or specialization of many farm types, every region still possesses the full range of farm types.

Because the patterns of farm type change are indeed so complex, many geographers have been content to classify agricultural farm types for one date only, or even to rely solely on officially published classifications; either of the type used by Bowler in the previous example, or the series of classifications produced by the Ministry of Agriculture for England and Wales (MAFF, 1980), this divides farms into the following 13 types of farming classes, as measured by their standard labour requirements with the key variables being 50 or 75 per cent of the standard man-days being devoted to the enterprise: (1) Specialist dairy; (2) Mainly dairy; (3) Livestock rearing and fattening, mostly cattle; (4) Livestock rearing and fattening, mostly sheep; (5) Livestock rearing and fattening, cattle and sheep; (6) Predominantly dairy; (7) Pigs and poultry; (8) Cropping, mostly cereals; (9) General cropping; (10) Predominantly vegetables; (11) Predominantly fruit; (12) General horticulture; and (13) Mixed.

Such classifications are however usually aspatial and based on solely economic criteria. Accordingly, geographers have attempted, and still continue

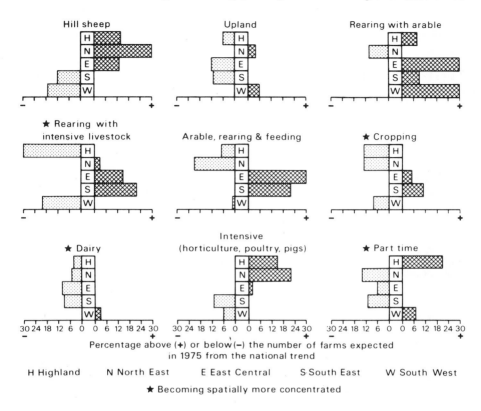

Fig 2.2 Regional changes in type of farming in Scotland 1968–75. Source: Bowler (1981)

to attempt, to produce more comprehensive classifications for the world, continental, country and regional scales (Kostrowicki, 1979; 1982). One of the first attempts was made by Weaver (1954), who argued that the most important feature of agricultural distributions was that crops are grown in combination with one another and not in isolation, and that terms such as the 'Corn' or 'Cotton Belt' were overgeneralized and oversimplified. Accordingly Weaver developed a method for defining crop-combination regions based on the percentage of the total harvested cropland occupied by individual crops. Weaver worked out an ideal model for one to ten crop-combination regions and compared this model to the actual crop combinations found in 1,081 middle-west counties, as shown in Table 2.3. From the table it can be seen that the deviation of the actual percentage of crop combinations in one county from the theoretical curve is lowest for a three-crop combination. Because the method in Weaver's (1954, 200) words; 'pertains only to crops and only to the land-use association of crops' Weaver concluded that the crop-combination regions should not be thought of as substitute agricultural regions.

Despite Weaver's own statement on the method's limitations and criticisms of his approach by Chisholm (1964), who concludes that his methods neither provide any more understanding of the regional relationships between crops and livestock, or whether it is corn or hogs that provide the main source of

Table 2.3: Weaver's method of producing crop-combination regions

Type of crop*	Monoculture C	2 Crops C	O	3 Crops C	O	H	4 Crops C	O	H	S	5 Crops C	O	H	S	W
% of cropland occupied	54	54	24	54	24	13	54	24	13	5	54	24	13	5	2
% theoretical base curve	100	50	50	33⅓	33⅓	33⅓	25	25	25	25	20	20	20	20	20
Difference	46	4	26	20⅔	9⅓	20⅓	29	1	12	20	34	4	7	15	18
Difference squared	2116	16	676	427	87	413	841	1	144	400	1156	16	49	225	324
Sum of squared differences	2116	692		927			1386				1770				
Sum divided by number of crops	2116	346		309			347				354				

*C = Corn (Maize); O = Oats; H = Hay; S = Soybeans; W = Wheat.
Source: Weaver (1954).

income, Weaver's method has been adopted and modified by a number of other workers, most notably Coppock (1964b).

Coppock also introduced another new element into the equation. The use of the computer to produce a hitherto unbelievable variety of maps based on 1958 agricultural census data, in his 'Agricultural Atlas of England and Wales' (Coppock, 1964a), and though in the first edition the computer-derived maps were redrawn by hand into traditional choropleth maps, by the time of the second edition based on 1970 census data (Coppock, 1976a) the computer maps themselves were often used, as shown in Figures 2.4 and 2.5. Both maps use the concept of man-days as the key variable behind the classification, but the mode of presentation is quite different between the two editions. The 1964 edition uses the Weaver method of presentation, but the 1976 edition presents a different map for the leading, second-ranking, third-ranking and fourth-ranking combinations respectively, as partly shown in Figure 2.5, and this makes them much more difficult to interpret. The second edition however also presents maps of certain types of farm as a percentage of all full-time farms, again based on the concept of man-days.

Although man-days provides a better parameter than land use, or sale of farm products, it must still be remembered that the resulting classification is based on aggregate parish data and cannot therefore be considered a type of farming map. The only such maps available are the 'Type of farming' maps produced by the Ministry of Agriculture and published either for the eight Ministry regions at the scale of 1:250,000, or for the whole of England and Wales at 1:625,000. In these maps the types of farm are determined according to the method used to produce the farm classifications described on page 22 (MAFF, 1980). Thus holdings are classified into a particular farm type if over 50 per cent of their standard labour requirements are in the enterprise of that type, or in the case of the predominantly dairying category, 75 per cent. Holdings with no enterprise above 50 per cent are classified as mixed. The 13-class classification is then simplified into nine classes and each holding is then located randomly in the parish in which it is recorded, and different coloured and sized dots are used to depict the holding, depending on its classification and size, as measured by acres on the series 1 maps or by standard man-days on the series 2 maps.

Nonetheless, Coppock's 1964 and 1976 maps of enterprise combinations (Figure 2.4) and leading enterprises (Figure 2.5) may also be just as typical of the type of farming found on the great majority of farms, although even the man-

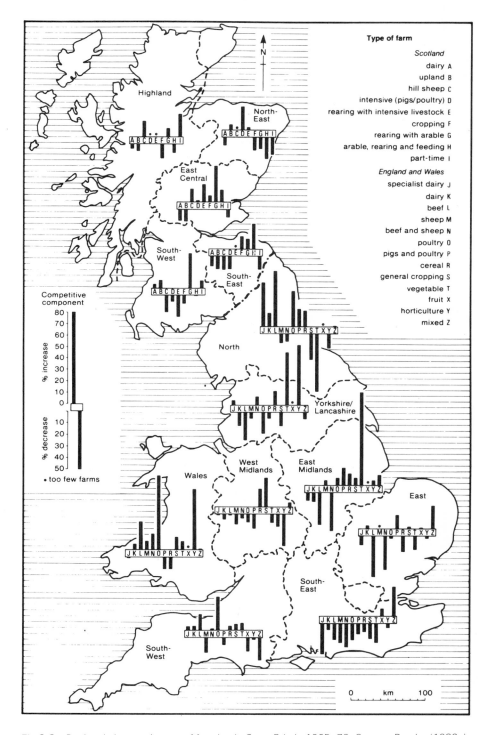

Fig 2.3 Regional changes in type of farming in Great Britain 1965–73. Source: Bowler (1983a)

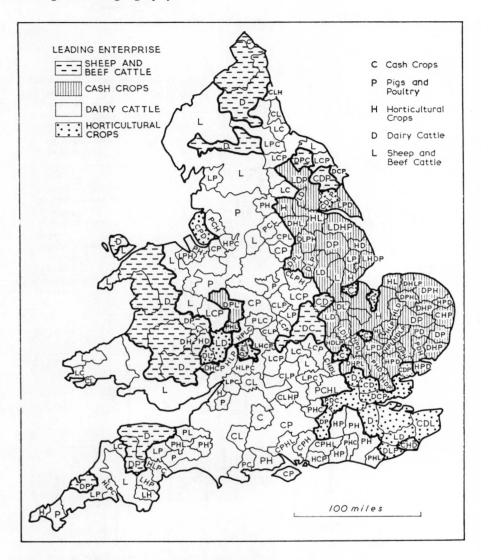

Fig 2.4 Agriculture enterprise combinations in England and Wales 1958. Leading enterprises shown by shading, secondary and other enterprises shown in rank order by letters. Source: Coppock (1964a) Reproduced with the permission of Faber & Faber.

day approach suffers from major difficulties over interpreting whether crops are grown for sale or for feeding to livestock, and whether cattle are beef or dairy cattle.

Therefore it is not surprising that Chisholm (1964) has laid down particularly stringent requirements for producing a type of farming classification system, in which the individual units classified must be the farms themselves, and where the classification must include all variables concerned with the production and management òf the farm, as well as information on yields, crops and livestock. Using these criteria, the map produced by Church and his colleagues (Church *et al.*, 1968), since it is based on agricultural census data alone, is invalid. Nonetheless, their map has one important advantage over other attempts since the authors, as civil servants, were able to use individual farm returns, and not the aggregated parish returns. Each farm was classified according to the relative proportions of standard man-days devoted to different enterprises and then assigned to one of five enterprises: arable cropping; horticulture; dairying ; beef cattle and sheep; and pigs and poultry, if over half their standard man-day inputs were devoted to that enterprise, if not they were allocated to the mixed category. Since the sample covered one sixth of all farms, namely a 50 per cent sample of farms within a sample of one third of parishes the resulting map is very representative, but still doesn't meet all Chisholm's requirements.

It is perhaps because of these rigid requirements that interest in classification procedures, apart from Coppock's two most useful atlases, appeared to wane in the 1970s. However, a few more contributions have been made. For example Munton and Norris (1969) examined 218 farms throughout England and subjected 36 structural and physical variables to a Principal Components Analysis and concluded that it had proved to be a valuable tool for data simplification and consequent hypothesis generation. In another example K. Anderson (1975) argued against Chisholm's (1964) view that (a) a very considerable degree of generalization of the data is required to make a comprehensible map of the whole country, and that (b) while this may serve admirably for descriptive purposes, it does not admit of further research. Anderson (1975, 148) in contrast argued that classification:

As a purely descriptive tool, . . . is extremely useful. When appropriately used, it provides a means of presenting at one time the significant features of a multivariate distribution. . . . Further, it permits the compilation of statistics in a manner which is more likely to reveal significant patterns. . . . A more important reason for classification is that it can stimulate further research, despite Chisholm's remarks to the contrary. Classifications can be extremely fruitful in the production of hypotheses.

Accordingly, Anderson used many of the statistical techniques developed since Weaver's 1954 article and in particular, hierarchical classification, which requires the computation of a measure of similarity or dissimilarity between every pair of objects in the initial data set. The two most similar objects are combined and the similarities for the new 'object' with the others are recomputed. The process continues in this manner until all the objects have been combined into a single group. Finally, discriminant analysis is applied to the set of optimized groups to produce a map of 11 enterprises for England and Wales for the 374 Agricultural Development and Advisory Districts existing in 1970 which, Anderson believes, reveals the considerable complexity of the spatial structure of agriculture.

In another example, Ilbery (1981) used a similar approach to Anderson and employed the four methodological stages shown below to produce an agglomerative multivariate classification of Dorset agriculture based on 35 variables derived from the 235 parish returns for the 1977 Dorset agricultural census:

1   Analysis of interrelationships among the variables used in the classification, by means of correlation coefficients.
2   Principal component analysis of the variables, to ensure that they have been transformed into Euclidean space and are orthogonal.
3   Cluster analysis, using Ward's algorithm, to group the observations (parishes) on the basis of the component scores.
4   Discriminant analysis for allocating marginal observations to the correct groups.

Analysis of the results suggested a more complex pattern than revealed by earlier studies (Tavener, 1952) and Ilbery concluded that such classifications could be used as a springboard for more detailed micro-studies.

This emphasis on using classification as a basis for further research advances has also been advocated by G. Clark (1984). He also made a distinction between two approaches to regionalizing, the inductive and deductive. The inductive regions which are created in an exploratory inductive fashion, are no more than mental artefacts and Clark argued that they are atheoretical and of little practical value due to their arbitrary nature. Conversely, deductive farming regions created to test a hypothesis, are easier to interpret and have a clear purpose in helping to advance geographical theory.

However, nearly all the techniques discussed so far have relied on secondary data collected for another purpose, namely the national direction of agriculture as part of the food industry. It is therefore not surprising that other approaches to the study of agricultural geography have been tried and in particular analyses of structure and employment, the subject of the next section.

Structure and employment

Although farm structure and employment appear on the surface to be less spatially relevant than regional differences in farm type one must agree with Bowler (1983b, 46) that farm structure – the size and spatial disposition of land holding – must be viewed as a fundamental factor in agricultural production, since:

> From it stem economies or diseconomies of scale in the use of labour and machinery; often it controls the availability of capital for investment in existing or new forms of agricultural production; increasingly, economies external to the farm, for example in the marketing of produce or the purchase of farm supplies, are denied to the occupiers of small or fragmented farms; above all farm structure is a major determinant of farm income.

Most analyses of farm structure have to rely on the agriculture census for their main source of data, but this is far from ideal. Firstly, the census is based very largely on physical information and has eschewed direct questions on the details of farm tenure and farmer's financial transactions (Ashton and Cracknell 1961). Secondly, about half the holdings listed are not farms in the sense of commercial full-time farms, but are part-time or hobby farms and this makes

Table 2.4: Distribution of holdings and standard man-days by type of farming

Type of farming	Number of holdings ('000)		Percentage of holdings		Percentage of standard man-days	
	1969	1975	1969	1975	1969	1975
England and Wales						
Dairy	52.0	42.9	39	36	30	30
Livestock	21.6	26.2	16	22	11	14
Pigs and poultry	9.0	8.8	7	7	9	12
Cropping	24.1	20.8	18	17	23	21
Horticulture	13.0	13.1	10	11	16	15
Mixed	13.4	8.0	10	7	12	8
Total	133.0	119.8	100	100	100	100
Scotland						
Hill sheep	1.4	1.0	6	5	6	4
Upland	4.8	6.0	21	31	16	23
Rearing with arable	3.6	3.4	16	17	11	13
Rearing with intensive livestock	0.6	0.4	3	2	3	2
Arable rearing and feeding	1.2	0.9	6	4	4	4
Cropping	3.7	2.9	17	15	21	18
Dairy	5.5	3.9	25	20	30	26
Intensive	1.4	1.1	6	6	9	10
Total	22.2	19.6	100	100	100	100

Source: MAFF (1977).

the distinction between a holding and a full-time business quite crucial (Ashton and Cracknell, 1961). A third problem relates to multiple occupation with either several farms or farmers working together in the same geographical area or running two separate units as a linked farm. This problem arises because the statistics have never been concerned much with farms as individual production units but are collected to produce overall production patterns (Commission of the EC, 1981).

To overcome some of these problems farms are often classified by the number of standard man-days needed, with 275 being the crucial limit for a full-time farm, or by their size. Indeed the census which used to use one acre (0.40 hectares) as its lower limit increased this to 10 acres (4.0 hectares) in the mid-1970s. although they still allowed very small farms, even below 0.40 hectares, to be included if they had a significant agricultural output, for example pig and poultry farms. Nonetheless, a further 15,000 statistically insignificant holdings were deleted in 1980 (Gilg, 1981).

In spite of difficulties with the data it is still possible to make useful generalizations about the changing structure of agriculture (MAFF, 1977), although the only regional breakdown for geographers is by country, as shown in Table 2.4. This reveals that the number of holdings fell by about 2 per cent a year between 1969 and 1975, but that within enterprise types some enterprises increased both their share and their numbers, notably 'livestock' farming in England and Wales and 'upland' farming in Scotland. Disregarding change it can also be seen that British agriculture is dominated by livestock farming, which accounted for some 65 per cent of holdings and 56 per cent of man-days in

Table 2.5: Structure and distribution of holdings by size and enterprise 1975

England and Wales	Size group (smd): 4,200+	1,200–4,199	600–1,199	275–599	<275
Number of holdings	3,870	28,250	42,270	45,150	86,360
Percentage of holdings	2	14	21	22	42
Percentage of smd	21	37	24	13	5
Dairy cows %	9	40	33	15	3
Beef breeding cows %	7	31	29	21	12
Breeding ewes %	6	32	34	20	8
Breeding pigs %	21	41	22	11	5
Laying fowls %	43	34	13	6	4
Broilers %	69	25	4	1	—
Wheat %	25	47	18	7	2
Barley %	16	45	24	12	5
Potatoes %	29	43	17	8	3
Sugar beet %	30	44	15	8	2

Source: MAFF (1977).

England and Wales in 1975. Since 1975 however there has been a transfer from dairy to cropping, as shown in Table 2.1.

There have also been changes within the size structure of holdings and the tenure pattern of holdings. The Ministry of Agriculture divides holdings into four sizes by standard man-days (smd): large (1,200 or more smd); medium-sized (600–1,199 smd); small (275–599 smd); and very small (under 275 smd) (MAFF, 1977). In 1968 in the United Kingdom there were 35,000 large holdings averaging 138 hectares and though they accounted for only 11 per cent of all holdings, they contributed 47 per cent of output as measured by smds. By 1975 their number had risen to 40,000 (15 per cent) averaging 143 hectares and they contributed 56 per cent of output. They had also continued to concentrate on certain enterprises, notably crop production, and eggs and poultry. In 1968 there were 58,000 medium-sized holdings, averaging 58 hectares, accounting for 19 per cent of holdings and 25 per cent of output as measured by smds. By 1975 although their number had fallen to 53,000 their average size, percentage share and output had hardly altered and they continued to concentrate their output in dairy and livestock production. In 1968 there were 87,000 small holdings but this total had fallen to only 65,000 in 1975 and although this still accounts for 25 per cent of all holdings their output had fallen from 20 per cent to 14 per cent, and their average size had only slightly risen to 31 hectares. Very small holdings have been affected by statistical changes in definition but it appears in 1975 that they numbered 115,000 or 42 per cent of all holdings, their average size was only 12 hectares and they contributed only 6 per cent of output. The more detailed pattern for England and Wales alone is shown in Table 2.5.

The tenure pattern of British farming has also undergone a marked change, as Table 2.6 shows, and as Table 2.1 reveals, this process has continued into the 1980s. In Scotland however, the percentage rented in 1975 remained at the same 41 per cent it was in 1961 (MAFF, 1977). In summary, in spite of the changes outlined above, British farming is still dominated in terms of numbers of businesses by smaller units, family ownership and management are the rule, companies are still relatively few, and the great majority of farms are run as proprietorships or partnerships (Commission of the EC, 1981).

Table 2.6: Area and numbers of holdings rented or owned 1960-75

England and Wales	Area			Holdings		
	Rented '000 hectares	Owned '000 hectares	Percentage Rented	Number Rented	Number Owned	Percentage Rented
1960	5,847	5,639	50.9	132,750	166,430	44.4
1969	5,148	5,945	46.4	96,420	149,790	39.2
1975	5,106	5,912	46.3	77,670	128,500	37.7

Source: MAFF (1977).

Nonetheless the structural changes of the 1970s have still had two main implications for rural areas. The first is that as the number of farms has decreased, and in particular the number of rented farms has decreased, there is less and less opportunity for people to enter farming as farmers, unless they are farmers' offspring. Accordingly, farming is becoming an increasingly closed occupation. The second implication is that fewer and fewer farmers have increased power over the countryside, not only in what they grow and produce there, but in how they manage the landscape, and as Chapters 3 and 9 demonstrate, the actions of this declining minority have been increasingly questioned by the growing majority of non-farming rural residents. Furthermore, as tenants are replaced by owner-occupiers, there is more freedom and incentive for farmers to act independently, and this is one reason why recent studies have emphasized the need to study decision making in agriculture.

It is indeed for this reason, and the aspatial nature of much of the census data discussed so far, that rural geographers have had to make more detailed studies of farming systems on the ground if they are to understand the processes behind the changes outlined so far. Unfortunately, the difference in scale between these studies and the national census based studies makes it very difficult to compare them and come to valid conclusions.

For example, Edwards's (1980) study of central Somerset covered only 400 square kilometres. Furthermore, the 25 per cent random sample of farms over 2 hectares, stratified into four sizes, produced only 122 sufficiently accurate questionnaires out of a total sample of 204 farms to give a 15 per cent sample. In spite of the difficulty of this sort of work, Edwards's interviews of farmers in 1963 and 1976 revealed some useful information, particularly the complexity of enterprises being operated, even in an area that could be considered as a core dairy area, as shown in Figure 2.4. In addition, as Tables 2.7 and 2.8 show, there had been a considerable change in the number and type of enterprises between 1963 and 1976, with an increasing concentration into single enterprises and a marked reduction in the number of farms running four or more enterprises. At the same time dairying lost ground as a first enterprise although it gained ground as the only enterprise. In conclusion, the study certainly showed the complex pattern of farm production systems and the degree to which even a so-called core area can change in only a few years as farmers 'respond dynamically, continuously changing the emphasis of their production systems, in response to temporal changes in the factors of production' (Edwards, 1980, 51).

Other geographers have studied structural factors at a larger scale. For example, Aitchison and Aubrey (1982) examined part-time farming in Wales on the basis of four key attributes: the scale of operation (by area); the commitment of the farmers (by numbers of hours per week); dependency (by farm income as

Table 2.7: Change in the number of enterprises in central Somerset 1963–1976

Number of enterprises	1963		1976	
	Number of farms	Percentage of sample	Number of farms	Percentage of sample
1	13	10.6	26	21.4
2	37	30.3	34	27.9
3	37	30.3	27	22.1
4	22	18.0	12	9.8
5	8	7.4	6	4.9
6	4	2.6	—	—
7	1	0.8	—	—
Retired	—	—	17	13.9
Total	122	100	122	100

Source: Edwards (1980).

Table 2.8: Change in the type of enterprise in central Somerset 1963–1976

Enterprise	Total frequency of occurrence		Percentage of sample		Percentage frequency of occurrence as:					
					Only enterprise		First enterprise		Subsidiary enterprise	
	1963	1976	1963	1976	1963	1976	1963	1976	1963	1976
Cereals	39	33	32.0	31.4	—	—	1.7	3.8	30.3	27.6
Potatoes	12	6	9.8	5.7	—	—	—	—	9.8	5.7
Dairy	112	85	91.8	81.0	9.0	20.0	80.3	52.4	2.5	8.6
Beef	22	53	18.0	50.5	—	2.9	1.7	10.5	16.3	37.1
Sheep	35	30	28.7	28.6	—	—	0.8	—	27.9	28.6
Pigs	67	29	54.9	27.6	0.8	1.9	4.9	4.8	49.2	20.9
Poultry	70	17	57.4	16.2	0.8	—	—	3.7	56.6	12.5

Source: Edwards (1980).

a percentage of total income); and career context (by previous occupation). Three areas were selected for detailed study, on the basis that the agricultural census revealed them to have higher than average incidences of part-time farmers. Within these areas, a questionnaire survey of 211 part-time farmers was undertaken, and most of the data collected from this survey were subjected to a cluster analysis, of the minimum variance variety, which produced six clusters as follows:

Cluster 1: 18 members. Most of these had moved from full-time farming to part-time farming and had developed some off-farm income.

Cluster 2: 26 members. All of these had previously farmed full-time, but had had to develop a much greater non-farm income than Cluster 1.

Clusters 3 and 4: 24 and 49 members. They have always farmed part-time and received off-farm incomes. Cluster 3 farmers derived their income on a 50:50 basis, but Cluster 4 farmers derived the great majority of their income from off-farm activites.

Cluster 5: 23 members. None of these had any previous experience of farming, but most of them regarded part-time farming as a step towards becoming full-time farmers.

Cluster 6: 71 members. This the largest group regarded the farm as a residence and a recreation and were truly part-time or hobby farmers.

However, although part-time farmers still account for a high percentage of holdings as shown in Table 2.5, an increasing percentage of output is provided by the small number of larger farms, since farm efficiency, however measured, indicates that farms are most efficient above 60 hectares in size, and between 600 and 1,000 smd (Britton and Hill, 1975). It is not surprising therefore that many countries including Britain have introduced schemes to rationalize farm structure, usually by encouraging small farmers to give up, but sometimes by imposing an upper limit on farm size either because of ideological grounds, as for example in Poland, or also because of suspected diseconomies of scale as farms become very large, as in France.

Three sorts of scheme have been identified in Europe by Hirsch and Maunder (1978): discontinuation schemes; farm enlargement schemes; and amalgamations or group farming. Bowler (1983b) has however argued that most evaluations of these schemes have emphasized the limited results that have been achieved, largely because of their voluntary nature and their inability to purchase land directly. Further difficulties are the short time span of most of the schemes, and the reluctance of farmers, however poor, to retire.

The main reason for this reluctance is the poor employment prospects that farmers face if they give up farming, and although rural employment is considered in more detail in Chapter 5, the employment situation in agriculture is perhaps best discussed here. It has already been pointed out that, as the number of holdings falls, and in particular, the number of rented farms is reduced, young people find it very difficult to become farmers, unless they are farmers' sons. The situation is exacerbated by the fact that the great majority of small farmers reject the idea of giving up farming to take another job (Gasson, 1969), either because their age and lack of experience of other work forms a barrier, or because they place a high value on the non-material advantages of the farming way of life. As a broad generalization it appears to Gasson (1969) that those small farmers who could move out of farming would not, while those who would move out of farming could not.

Nonetheless, farming in the developed world is becoming increasingly dominated by large farms employing fewer and fewer workers, leading to fears of an unbalanced social structure, even in America, where the concept of a property-owning democracy of small farmers, which is so vital to the American dream of society, is under threat (Goldschmidt, 1978).

In Britain, the agricultural labour force employed by farmers has declined for over a century, and although some 50 per cent of the world's population was still employed by agriculture in 1975 (Grigg, 1976) the British figure had fallen to less than 3 per cent of the labour force by 1982 (MAFF, 1983a). There are probably less than 200,000 full-time workers left and over 75 per cent of these are employed on large farms employing four or more workers. Over most of the country there are fewer than 15 regular workers per 400 hectares of agricultural land, and only in the horticultural areas around London, in Kent, the Fens, and the south coast between Bournemouth and Brighton does the figure rise to over 25 workers per 400 hectares (Coppock, 1976a).

The main reason for the loss of farm work has been mechanization, but there is some dispute about the exact process of loss (Wagstaff, 1974) and although Gasson (1974) has demonstrated that 47 per cent of workers left of their own accord, for the variety of reasons shown in Table 5.7, her data refer only to East Anglia and the period 1960–62. A more recent study (Errington, 1980) has shown that the agricultural labour force is more complex than often thought, but that dissatisfaction with the large social and economic gap between farm worker and farmer was one major reason for workers leaving farm employment. However, Newby's (1977) major study of farmworkers in East Anglia revealed that though the objective deprivation of the farm worker is probably as great today as it has ever been, he nonetheless displays few signs of dissatisfaction. However, because of his relative powerlessness to obtain higher rewards within farming, owing to the still quite rigid constraints that operate in both the labour and housing markets, his only choice is to leave the land altogether. In conclusion, the choices made by both farm workers and farmers are now seen to be fundamental in any explanation of agricultural geography and so accordingly, attention is now directed towards decision-making in agriculture.

Decision making in agriculture

The growing interest in decision making in agricultural geography reflects the wider development of rural geography outlined in Chapter 1 from normative to behavioural studies. This interest is not just the natural evolution of the three stages of agricultural study proposed by Tarrant (1974), namely: physical determinism in which the physical environment controls agricultural decision making (1900–55); economic determinism in which farmers react in a uniform and rational manner to economic circumstances (1955–75); and behaviouralism in which a further set of influences which are neither physical or economic are thought to influence farmers' decision-making processes. It is also the result of a shift in direction away from studying form – for example, farm type regions or farm structure – to an analysis of the underlying processes producing these forms.

This chapter has already shown that land use patterns are not only very complex, but also very difficult to analyse and explain, largely because they are produced by the managerial decisions of thousands of individual farmers who are seldom motivated by economic considerations alone. Accordingly as long ago as 1966 Harvey, in a review article, argued that normative theories of agricultural location patterns should be adapted to take account of psychological and sociological realities, and Wolpert (1964) working empirically in Sweden found that the average farmer only achieved two-thirds of the potential productivity which his resources would allow, and he therefore assumed, and his subsequent analysis demonstrated, that one or both of the prerequisites for economic rationality (perfect knowledge and optimizing behaviour) were absent.

Ilbery (1978) has summarized the literature on decision making which has developed from Harvey's and Wolpert's early ideas and has concluded that there are three aspects to the topic. First, the analysis of agricultural objectives, centred on the aims of an individual and why he acts as he does. Second, the context in which a decision is made, for example, a farmer may well act on personal feelings if hard information is difficult or expensive to obtain, or on the experience of neighbours following the theory of agricultural innovation developed by Hägerstrand (1967). Third, the methods employed in arriving at a course of action will depend on a number of factors, for example, the amount of time available or the age of the farmer. These three aspects have also been recognized by agricultural economists and texts are now available (Anderson *et al.*, 1977) which aim to guide farmers towards better decisions by the use of decision analysis which involves the definition of relevant acts and a statement of their consequences; the assessment of possibilities and the seeking of further information; and then the selection of the optimal strategy on the basis of max-imizing expected utility.

Within Ilbery's three contexts, two main approaches to behavioural model-building have been developed, first, empirical and normative decision-making models, and second, satisficer versus optimizer concepts. In the first case, either an empirical model is set up to discover patterns and regularities, and then deviations from this are described, or a normative theory is set up which asks how a rational decision maker might act in a given situation. In contrast to these optimizer models which assume objective reality, the satisficer concept argues that farmers will not or cannot examine all the possible solutions, but will embrace the first satisfactory solution they encounter. This model therefore assumes a stochastic or random element, but this unfortunately is very difficult to build into a model convincingly.

Whichever type of model is used, research workers have tended to examine either the flow of information to and between farmers, or the socio-personal factors affecting their decisions. Hägerstrand (1967)developed both a stochastic and an information-flow approach in his classical development of diffusion theory in the early 1950s. The basis of the theory is that in the early stages of an innovation, a few farmers who are particularly 'go-ahead' and energetic, actively seek out new techniques and are early adopters of innovations. How-ever, most farmers are conservative and only adopt the innovation later on after they have seen their neighbours using it successfully. The process can be summed up as an 'S-shaped' curve, in which a few people only adopt the innovation initially. This is followed by a rapid uptake by the majority, and then by a slow adoption as the few remaining laggards take up the innovation at the end of the period.

Other workers have examined the wider factors influencing a farmer's decision-making process. For example, Fotheringham and Reeds (1979) in their survey of tobacco farmers in Southern Ontario used discriminant analysis to find that four variables were most important in explaining a farmer's choice of suitable crop, namely age, area of farm, amount of labour locally derived, and the amount of expansion recently undertaken. Taking a more social than economic view Gasson (1973) in her study of East Anglian farmers found, as shown in Table 2.9, that 'intrinsic' values were far more important than 'expressive', 'instrumental' or 'social' values. This reinforces earlier points that farmers place a high value on either 'independence' or 'way of life' and are thus less likely to act economically and rationally than, say, an urban employer or industrialist.

Gasson's work in East Anglia was followed 10 years later by a very similar study of West Midlands hop farmers by Ilbery and although the results shown in Table 2.10 weren't completely compatible, Ilbery confirmed that the hop farmers also placed more importance upon doing the work they like and being independent than on the income aspects of farming. The three main dis-

Table 2.9: Attractions of farming life suggested by East Anglian farmers

Attribute		Size of farm business	
		Under 600 smd	Above 600 smd
		Per cent of responses	
Independence	Intrinsic	42.8	13.5
Way of life		27.0	32.2
Aspects of work itself		14.8	8.5
Challenge, risk, gamble	Expressive	5.0	1.7
Achievement, creativity		2.7	11.8
Pride of ownership		0.0	3.4
Self-respect		0.0	1.7
Income, cheaper living	Instrumental	3.2	6.8
Chance of capital gain		2.3	5.1
Family tradition	Social	0.9	8.5
Interaction with other farmers		1.3	6.8
Intrinsic		84.6	54.2
Expressive		7.7	18.6
Instrumental		5.5	11.9
Social		2.2	15.3

Source: Gasson (1973).

Table 2.10: Goals and values of East Anglian farmers and West Midlands hop farmers

Attributes of farming occupation	Rank	East Anglia	West Midlands	Rank
Doing the work you like (I)	1	174	342	1
Independence (I)	2	145	340	2
Making a reasonable living (In)	3	135	335	3
Meeting a challenge (E)	4	129	317	6
Leading a healthy outdoor life (I)	5	110	313	7
Expanding the business (In)	6	100	220	17
Making sure of income for future (In)	7	93	295	9
Being creative (E)	8	91	268	12
Making as high an income as possible (In)	9	86	281	10
Working close to home and family (S)	10	84	266	13
Self-respect for doing a worthwhile job (E)	11	84	323	5
Earning respect of workers (S)	12	81	329	4
Following in the family tradition (S)	13	76	196	19
Being able to arrange hours of work (In)	14	71	218	18
Job security+/control*	15	67	302	8
Belonging to the farming community (S)	16	55	238	16

I = Instrinsic. E = Expressive. In = Instrumental. S = Social.
+ Gasson's term. * Ilbery's term.
Sources: Gasson (1973) and Ilbery (1983b).

crepancies: earning the respect of workers, self-respect for doing a worthwhile job, and expanding the business, were partly explained according to Ilbery by the nature of hop farming. However, Ilbery could find no significant difference between hop farmers in terms of farm size, as revealed by Gasson's work in Table 2.9, and suggested that these differences may be due instead to type of farming. Ilbery also produced a typology of hop farmers based on 32 variables (16 of which are listed in Table 2.10), broken down into 20 attributes based on farming occupation and 12 based on farm and farmer characteristics, for example , farm size, education and age. Using principal components analysis he also produced a simple divison into two types of hop farmer, and by using cluster analysis, he also produced a typology of nine types of hop farmer.

Similar methods were also used by Ilbery (1979) in his study of 102 farmers of all types in northeast Oxfordshire. Using principal components analysis, cluster analysis and point score analysis Ilbery (1977) was able to produce the results shown in Table 2.11, which led him to conclude that the main aim of the farmers was to obtain a secure and stable farm business. Once the essential factors for the achievement of security were achieved, farmers were more strongly influenced by social and personal considerations and they chose to work with an enterprise that gave them most satisfaction.

However, as Gasson's work (Table 2.9) has already shown, decision making alters with farm size and so Ilbery also subdivided his sample, by the use of cluster analysis, into a typology of eight groups, ranging from the largest group of 35 full-time mainly small-scale farmers who were most motivated by market/demand, income and free time to 12 full-time mainly young farmers who had obtained good practical training and considered free time an irrelevant factor. Another typology has been developed by Newby *et al.* (1978) based on

Table 2.11: Ranking of agricultural decision making factors by order of importance

	Total score	Rank
Economic factors		
Stable market/demand	642	1
Regular income	524	2
Above average profits	523	3
Did not require much labour	332	8
Did not require much working capital	213	10
Available buildings/machinery	164	12
Transport costs	138	13
Government policy	74	16
Under-used land available	56	18
Co-operatives	33	19
Socio-personal factors		
Experience	490	4
Personal risk	479	5
Free time	385	6
Personal preference	363	7
Proven type of farming in area	314	9
Enterprise already here	176	11
Trained staff	106	14
Agricultural training	82	15
Prior knowledge of enterprise	59	17

Source: Ilbery (1977).

Table 2.12: Newby's typology of East Anglian farmers

		Market orientation	
		Low	High
Degree of direct	Low	Gentleman farmer	Agri-businessman
involvement in husbandry	High	Family farmer	Active Managerial manager

Source: Newby *et al.* (1978).

the degree of direct involvement in husbandry and market orientation as shown in Table 2.12. Newby and his co-workers found that agri-business dominated the large farming scene with 54.3 per cent of farms above 400 hectares being farmed by agri-businessmen, but only 12.3 per cent of all farms being farmed by agri-businessmen. On this wider scale, family farms dominated and accounted for 56.1 per cent of all farms. While Newby himself admits that his samples are not representative, and the same could be said for both Ilbery's and Gasson's samples in that they are either based in central and eastern England or on certain types of farmer, they all throw a good deal of empirical and theoretical light onto the previously neglected area of decision making. Although more research clearly needs to be done, in particular into the extent to which farmers give the answers they feel they ought to give, as revealed by Ilbery and Hornby's (1983) preliminary work on repertory grids, it is still possible to generalize about agricultural decision-making processes (Blacksell and Gilg, 1981, 36) and to conclude that:

> The main factors affecting attitudes to change are age and education, whether a farmer owns or rents the farm, the desired level of income and the risk each individual is willing to take to achieve it, size of farm, the availability and cost of finance, the level and stability of government aid and price support, the rate of taxation, and, finally, the likelihood of the farm remaining in the family when he retires.

So far this chapter has examined three main approaches to agricultural geography: (a) the regional change, classification and regionalization approach; (b) the structural and employment approach; and (c) the decision-making approach. It is to be hoped that once the results of the behavioural (decision-making) approach are integrated with the other two approaches as advocated by P. Hart (1980), a more useful agricultural geography based on more generally applicable agricultural location theories should emerge, in contrast to the rather unconvincing agricultural geographies produced so far.

Agricultural geography and agricultural location theory

In the early 1970s books on agricultural geography appeared almost as frequently as the annual cycle of the farming season, but like the weather in each season, they were just as variable in their content and no one text was like another.

One of the longest-running texts, Symons (1978) which was first published in 1967, divides its material into the physical and social environment, systems of exploitation, and concepts and methodology. Symons therefore takes a traditional and world-scale view, with chapters on farming in Russia, Malaysia and in subsistence areas, though this coverage isn't as comprehensive as Grigg's

(1974) text on the agricultural systems of the world, and Symon's chapter on 'Systematic and Regional Analysis' only takes up 37 of the book's 285 pages. Four years later, in a text confined to Great Britain, Coppock (1971) took the traditional route of the agricultural economist and divided the subject material into the 'so-called' factors of production: land and weather, farms and fields, men and machines, markets and marketing, and land and livestock, before considering each production sector: dairying, beef cattle, sheep and lambs, pigs and poultry, crops and horticulture separately. In spite of using this approach Coppock argues that there is a radical difference between the two approaches of the agricultural geographer and the agricultural economist.

Basically the agricultural economist believes that intra-region differences between different farms are greater than the inter-region differences between different farming regions (Britton, 1968), and therefore that any search for a greater degree of spatial order is misguided. In contrast the agricultural geographer (Coppock, 1971, 9):

> retains a conviction that the distribution of farm-types and enterprises is less haphazard than economists suggest, that the regional differences in the factors of production that have been shown to exist must affect both the efficiency and profitability of farming, and that more sophisticated data would reveal smaller intra-class and wider inter-class differences than are now apparent from the use of large regions.

However, Coppock leaves the reader to construct his own regional agricultural geographies, since the book is highly systematic and only 22 of the book's 345 pages are given over to the 'pattern of farming'. Another example of the systematic approach based around the factors of production is provided by Edwards and Rogers (1974). In this edited book a combination of agricultural economists and geographers discuss structure, markets, technology, land, labour, capital, management, the pattern of farming, competition for farmland, land assessment and data sources, to provide a balanced picture of agriculture if not a comprehensive agricultural geography.

Indeed Morgan and Munton (1971, 1 and 3) argue that: 'To some extent a single discipline of agricultural geography is difficult to conceive' but that if it can be, that it 'is a part of economic geography.' Accordingly, their text adopts a systematically economic approach with chapters on enterprises and systems, the farm firm, agricultural activity and the physical environment, land, labour and capital, scale of production, marketing and supply, government policy, data and classification, and patterns of distribution.

Perhaps the most comprehensive agricultural geography yet produced is Gregor (1970), largely because it surveys research progress, and doesn't attempt to present a universally accepted body of information that doesn't really exist. Gregor divides his material into four main sub-divisions: the aims and methods of research; the study of landscape from an environmental, spatial, cultural, political or historical context; the search for regions; and the concern for resources.

The first book to really radically depart from these traditional approaches to the subject was Tarrant (1974), since it begins with a consideration of agricultural location theory, continues with chapters on data sources, regionalization and classification, statistical techniques, and marketing, before reverting to more traditional themes in a final chapter on the competition for agricultural land. Tarrant's book is therefore based on methodological and theoretical approaches to the subject, in common with Gregor's text, and

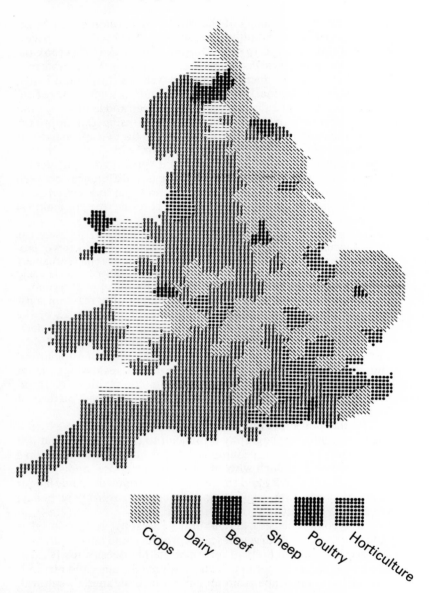

Fig 2.5 Leading and second ranking agricultural enterprises in enterprise combinations: England and Wales 1970. Leading enterprises on the left, second ranking enterprises on the right. Source: Coppock (1976a) Reproduced with the permission of Faber & Faber.

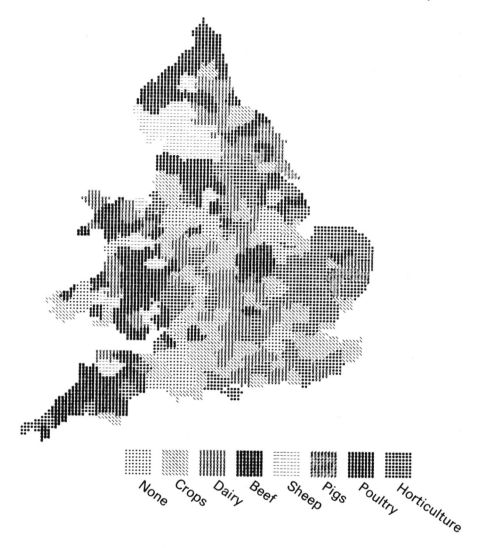

reflects the quantitative and methodological revolution of the 1960s, and perhaps most of all, the development of agricultural location theory.

Although agricultural locational theory (Visser, 1982) was first brought to the attention of a wide audience by Chisholm's (1962) book *Rural Settlement and Land Use* which examined the theoretical basis of Von Thünen's model and empirical evidence to test the theory, it wasn't really a creature of the 1960s since its history goes back to the Napoleonic Wars, when Ricardo proposed the concept of economic rent, which has remained central to ideas of agricultural location, particularly those of a deterministic character. Ricardo used two variables to account for variations in agricultural rent, namely soil quality and population density. In 1826 Von Thünen (1966) substituted distance for soil fertility in Ricardo's model, while at the same time allowing a greater diversity of agricultural produce. In Von Thünen's model therefore, transport costs are the cause, and rent the consequence, of differences in land use. In order to simplify the model, Von Thünen proposed a uniform plain and one market, a single central city, although he did later introduce complicating concepts, for example, the idea of a navigable river as shown in Figure 2.6, to produce the concentric rings of land use based on transport costs and declining rent gradients with distance from the market, so typical of the model. Although Von Thünen's model can be seen to operate in many circumstances, for example, the increasing practice of contracting crops to processing plants (Hart, 1978) and the less intensive use of fields with increasing distance from the farmstead (deLisle 1982; Found, 1970) and in many areas, notably in the less developed world, the complicating effects of international tarriff barriers and internal government price policies have distorted the effect of distance (Moran, 1979) even within theoretically tarriff-free areas like the EEC (Belding, 1981).

The most notable example of the effect of distortions in transport costs is provided by the shift of milk production from the cities in nineteenth-century Britain to the more physically suited areas of the west country in the twentieth century as transport costs were reduced, as a relative cost, as a deliberate act of policy (Strauss and·Churcher, 1967). In an American study of transport costs, Winsberg (1980) also found that reduced travel costs had increased the concentration and regional specialization of 19 farm commodities between 1939 and 1978 as measured by the location quotient and the index of dissimilarity, and that agricultural innovation, although reducing the impact of comparative advantage, had not led to a deconcentration of regional specialization. However, Visser's study of Great Plains agriculture (1980) revealed that the Von Thünen model proved to be more useful at the local rather than the regional level, since technical innovation distorted the regional pattern, for example, the uneven spatial introduction of irrigation. In contrast, Muller (1973) using trend surface analysis concluded that Thünian distance processes had played a dominant role in shaping the macro-geographical pattern of agricultural production in the United States.

Theoretically, Jones (1983) has (a) extended the model to include the production of agricultural goods for local consumption, (b) related farm tenure (Jones 1982), risk-related activities (Jones, 1984a), and off-farm employment (Jones, 1984b) to distance from the market, (c) argued that a useful expansion of the model would be·to make the price of a non-tradeable good, such as housing, endogenous (Jones, 1984c) and (d) has concluded that his work has reinforced the robustness, flexibility and usefulness of the Von Thünen model. In another alteration, Cromley (1982) has modified the model to take into account the probability that the crop pattern that produces the best short-term profit may not always produce the best long-term profit.

In spite of these alterations, both actual and proposed, the Von Thünen model still has a number of limitations, for example, the assumptions that each farmer has complete information, and that each operator makes rational decisions to maximize income. As the previous section on decision making has illustrated, this is not always the case, and so location theory has had to be modified to take into account behavioural considerations. This can be done normatively by the use of game theory or by empirical studies of the type outlined in the previous section.

Another limitation to the Von Thünen model is the uneven diffusion of

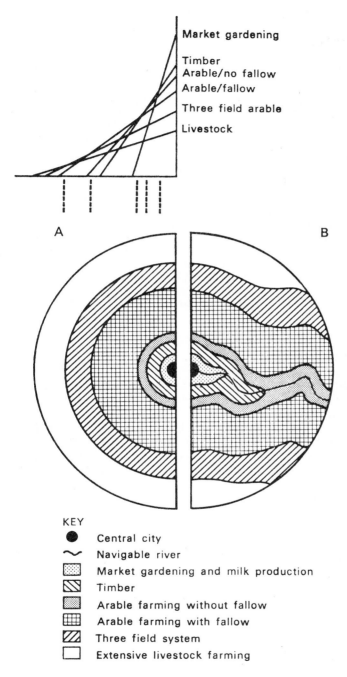

KEY
- ● Central city
- ∼ Navigable river
- ▦ Market gardening and milk production
- ◩ Timber
- ▦ Arable farming without fallow
- ▦ Arable farming with fallow
- ▨ Three field system
- ☐ Extensive livestock farming

Fig 2.6 The Von Thünen model of agricultural location. The top diagram shows rent curves declining with distance from the central city, and the lower diagram shows the same in spatial form, for both an isotropic plain (left), and with a navigable river (right). Source: Tarrant (1974)

innovations, and although Hägerstrand (1967) has developed a very useful model which appears to be reflected in practice more often than not, the actual distribution of farmers who are motivated by a number of very different factors and who may or may not react to an innovation, or a change in market price, or government policy, makes the formulation of any general model a very difficult task indeed.

Although written as long ago as the early 1970s, Found (1971) remains the best attempt to produce such a general model, by blending the approaches of the normative models of traditional economics to the more recent attempts to formulate concepts of man's psychological behaviour. It is a pity that no text on agricultural geography written along these lines apart from Huggett and Meyer's (1980) sixth-form text book has been produced since Tarrant's work in 1974. Furthermore, neither the development of further theory or its empirical testing has really occurred in the years since Found argued for both these developments in his closing paragraph. Indeed his closing sentences remain as valid for the mid-1980s as they were in 1971 (Found, 1971, 168):

> In many cases, theoretical models have been well developed, yet little empirical testing has occurred. Models of comparative advantage, of general spatial equilibrium and almost all behavioural formulations are examples. Numerous references are available on game theory, learning, decision environments and related concepts; but practically no testing of these concepts in the real world has occurred. Until such testing occurs our theoretical formulations and understanding of land-use decision making will be greatly impeded.

Concluding remarks

This chapter has presented a more logical and rational picture of agricultural geography than really exists. There is in reality no coherent agricultural geography, and indeed no new book on the topic has been produced since 1974, *although economic and systematic texts continue to be produced (Andreae 1981; Engledow and Amey, 1980; Newbury, 1980; Wormell, 1978). There are however a number of important themes within agricultural geography that are still being followed or further developed, for example, Coppock's continuing and invaluable mapping work (Coppock, 1976a; 1976b), Bowler's very useful work on the regional uptake of government grants (Bowler, 1976c; 1979), and Ilbery's innovative work on decision making (Ilbery, 1979). In a wider context, Grigg (1981; 1982; 1983) in his annual reviews of agricultural geography has suggested that though there have been no profound changes (1981, 271):

---

* As this book went to press one new text did however appear (Grigg, D.B. 1984: *An introduction to agricultural geography*. London: Hutchinson), and this provided the following definition (p.13): 'Agricultural geography seeks to describe and explain spatial variations in agricultural activity over the earth's surface'. And in greater detail, 'description' was divided into (a) the systematic analysis of one variable and (b) the production of type of farming maps and the definition of agricultural regions; and explanation was divided into (a) studies of the relationship between environment and agriculture, (b) locational and economic models and (c) behavioural approaches.

That interest in the Von Thünen models has reached the baroque stage; that diffusion studies are beginning to produce diminishing marginal returns; that interest in the institutional determinants of agricultural activity such as land tenure and government policy is increasing and there is even some evidence of a revival of interest in environmental influences in agriculture.

In another review Bowler (1984) has argued that four new issues have emerged to challenge the more traditional themes, namely the characteristics of modern farming systems, the loss of agricultural land (covered in Chapters 8 and 9), State intervention (covered in Chapter 9) and multiple-job holding or part-time farming.

It isn't easy to explain why agricultural geography hasn't developed any faster as a coherent discipline though there are probably three main reasons. The first is that a reasonable degree of description and explanation has been achieved, and that any further advances have to overcome the formidable data and methodological problems outlined earlier in this chapter. The second is that geographical studies have tended to narrow down from the macro to the micro-scale in the last few years, and this has militated against agricultural geography which has traditionally focused on at least the regional if not the world scale. The third reason is that interest in the rural physical environment has switched from farming *per se* to the effects of farming on the environment, in terms of pollution and landscape change and Simmons (1980) has proposed a radical new approach, based on viewing agriculture as a set of ecosystems in which energy flows and nutrient cycling are the main ecological models. Although these are very good reasons, it is nonetheless to be hoped that some geographers will return to the more traditional interests of regional agricultural geography while at the same time, others extend the theoretical and behavioural background.

3

Forestry, mining and land use competition

Agriculture is not the only extensive land use that is of interest to geographers but it remains the most researched and written about. Nonetheless there have also been a number of studies of forestry and mining, originally as topics in their own right, but increasingly as examples of land use competition and integration. Accordingly, this chapter examines first, the way in which forestry has partly changed from being regarded as a single-goal orientated industry, with economics being the main concern, to a multiple land-use activity, second, the growing conflicts over the extension of mineral extraction to rural areas, and third, the supreme example of rural land use competition, the urban fringe. In the process, this chapter also introduces some of the basic concepts of rural land management and planning that are discussed more fully in Chapter 9.

Forestry

Forestry is of interest to rural geographers first, because it is the second most extensive user of rural land after farming, second, because forestry competes with agriculture for the use of marginal land both in the lowlands and uplands, and third, because trees are such long-lived and large features in the rural landscape. Accordingly, any study of forestry must take into account the longevity of trees, and the fact that even the fastest-growing trees take from 40 to 150 years to reach any sort of maturity. This relative stability of forestry, compared to the more dynamic changes experienced in agriculture, may explain why only a few rural geographers involve themselves directly with forestry research (Watkins and Wheeler, 1981), although they often use forestry as a vehicle for the study of land use issues. Another reason, is that the industry is dominated by government agencies, for example, the Forestry Commission, and a few very large companies and landowners, and so doesn't provide the same scope for research into decision making or behaviouralism. Indeed most of the interest in forestry has been concerned with arguments for and against afforestation, and whether it should replace upland hill farming, or be seen as a complementary use of the uplands. These arguments can best be understood by taking an historical perspective, and seeing how the economic and strategic arguments have gradually been reassessed in the light of changing circumstances.

The economic and strategic arguments in favour of forestry reigned supreme throughout the first half of the century, and the results, in semi-mature timber, are there for all to see in the uplands of Scotland, Wales and northern England (Forestry Commission, 1967). Indeed the planting programmes of these subsequent years have caused one of the largest-ever landscape changes in the United Kingdom (Mather, 1978), and since the Forestry Commission was

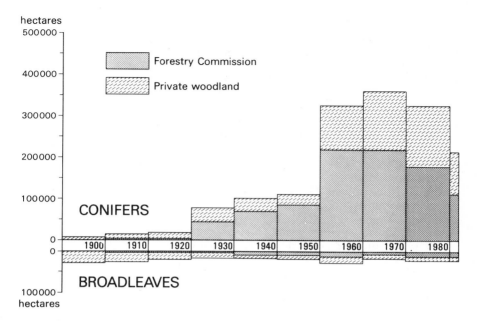

Fig 3.1 Area of new forest by decade of planting. Planting rates in the UK reached a peak in the 1950s and 1960s, since then the rate has fallen largely due to a drop in Forestry Commission planting. Source: Gilg (1984)

formed in 1919, 800,000 hectares have been planted (Price and Dale, 1982) as shown in Figure 3.1. The reasons for this rapid reafforestation of Britain stem largely from the depredations of timber in the First World War. This cruelly revealed the poor state of Britain's forests, which had been in decline since the late Middle Ages. Accordingly, in 1919 the government set up the Forestry Commission to plant new forests. However, before any of these new forests could reach maturity, the Second World War made even further demands on Britain's already severely depleted woodland stock, and so in 1943 a post-war afforestation policy set a target of two million hectares of forest by the turn of the century, which was to be achieved by Forestry Commission planting and by reinvigorating the private sector with grant aid and taxation concessions. Although it now looks unlikely that this target will be achieved, the basic aims of the Forestry Commission (1983a) and the forestry industry still relate to further economic expansion, although these goals have been increasingly questioned in recent years.

The first serious questioning of the expansion policy, came in the late 1950s, when the strategic arguments became weaker with the advent of nuclear weapons. Forestry was therefore given a social and employment role in addition to its economic function, and it was argued that forestry could provide more jobs (albeit at a higher cost than hill farming, see Table 5.10) and keep more people in remote rural areas than hill farming (Mather, 1978). These assumptions were however queried in the mid-1960s (DES, 1965) when an inter-disciplinary group concluded that agriculture and forestry could only be compared by discounting future expenditure and revenue to the initial year of the project. Using these criteria, the group argued that agriculture earned a higher return on marginal land than forestry, unless the discount rate was as low as 3 to 4 per cent, and that on better land, even a discount rate of 3 per cent couldn't equal the agricultural returns. This report, however not only undermined the economic case for forestry, it also gave more attention to alternative uses for the uplands and in particular to both the concept of multiple land use, and the integration of farm and forest land with two land uses that were already growing in importance; conservation and recreation.

The force of these arguments was not however fully tested until a 1972 cost/benefit study (Treasury, 1972) came to the conclusions listed in Table 3.1, and heralded not only the end of the expansion period, but also the beginning of the multiple land use period. Nonetheless, the case for forest expansion continues to be argued (Doyle, 1977) if less strongly than before and even the Forestry Commission (1977, 64) has been forced to be more cautious:

> It is clearly impossible to make any other than a subjective assessment of the overall impact of the different planting programmes considered (no planting from 1977–2025, 1 million hectares of planting, and 1.8 million hectares of planting). It must remain a matter of judgement how much weight should be given to the various factors. . . . Those such as reduced dependence on imports . . . may not be . . . so vital . . . when looking ahead 50 years, but the selective profitability of wood growing, environmental aspects and the provision of rural jobs seem likely to remain important issues. On balance the Group believe . . . the reasonable and prudent course . . . is to maintain and indeed increase the rate of planting.

A less cautious forecast was however provided by the Centre for Agricultural Strategy (1980) who argued that world shortages of timber would lead to real price increases of 150 per cent by 2025, and that this would make afforestation profitable at a range of discount rates. They therefore proposed the afforestation of a further two million hectares. Coleman (1980), in her call for a doubling of the woodland area, also used the growing pressure on world timber supplies as a basic argument, but instead of using price as the key variable, argued that self-sufficiency of timber as a renewable resource was the most important principle.

However, Price and Dale (1982) have argued that the rate of future price rises is still acutely critical to the afforestation case, and therefore the amount of land that could be said to be economically plantable. In a study in North Wales they found that with no price rises, only 19 per cent of the rough grazings studied were economically afforestable, but that with a 3.5 per cent per year price rise, 83 per cent was plantable.

Although the discount rate is clearly the most crucial financial variable in the forestry debate, forests are no longer seen as simply economic factors of production, providing timber and jobs, but also as places for recreation, important conservation sites and significant features in the landscape (Watkins and Wheeler, 1981). Even in the USA where 33 per cent of the land is still under trees, Clawson (1975a) has listed no less than seven pressing issues for forestry policy, emphasizing the complexity of the forestry debate.

The diversity of land uses that forestry can support is therefore a central feature of the forestry debate, and though the government has refused to give a clear lead (HL Select Committee, 1982) on how to integrate forestry, both internally and externally, with other uses, in spite of calls to do so from both Parliament (HL Select Committee on Science , 1980) and outside (Lovejoy, 1983),

Table 3.1: Conclusions of the 1972 Cost/Benefit study of forestry

1   There appear to be no strategic considerations relevant to new afforestation.
2   The recreational value of the estate is as yet modest but its future value is substantial. The recreational value of new planting is more limited, because of its location in remote areas and because the current estate will provide an extensive resource when it reaches maturity.
3   Hill-farming in the areas being afforested, North Wales, Southern Scotland and Northern Scotland produces a negligible or even negative social return at a 10% discount rate.
4   New planting fails to produce a 10% social rate of return and would fail to do so even if forestry were to maximize its regimes.
5   Re-stocking after clear felling can produce a 10% social rate of return.
6   New planting compares unfavourably with the hill-farming it replaces in economic resources and exchequer costs per hectare, but it creates slightly more employment albeit at a higher cost per job in resource and exchequer terms.

Source: Treasury (1972).

multiple land use on the ground has increasingly become a way of life. For example, a 1960s survey of 72 estates showed that 22 estates used their woodland not only as forestry but as shelter from strong winds (Nicholls, 1969), and Carroll (1978) was able to argue that the intrinsic characteristics of the woodland owner were more important than either economic models or environmental factors in explaining the degree of multiple use.

Multiple use can take a number of forms. *Economically*, it has been argued (Fraser, 1982) that the half a million hectares of mainly lowland deciduous woodland that are classified as commercially unproductive could be developed for craft work, charcoal burning and domestic heating, and Towler (1975) has shown how the integration of Forestry Commission planting and hill farming in North Wales has produced a 30 per cent increase in lamb yields. In the area of *recreation*, Tandy (1978) has described how forestry interests and recreation are now considered together in Forestry Commission planning although they had at one time been considered incompatible, since the walkers were a constraint on commercial timber operations, and the dark spruce plantations were considered an eyesore and a hindrance by the ramblers. However, although forests are now widely used for recreation, not everyone would agree with Tandy's view of the relationship, and there is still considerable opposition to further afforestation by a section of the rambling lobby (Ramblers Association, 1980). In the field of *conservation*, although a good deal of progress has been made in the scientific theory and actual practice of combining nature conservation and commercial timber operations in the modern forest (Peterken, 1981), the post-war period has been a time of unprecedented destruction of the ancient woodland heritage (Rackham, 1976). This long *landscape* heritage is also considered by Miles (1967), and fortunately, his arguments in favour of landscaping new woodland into the landscape have by and large been followed by the Forestry Commission, who since 1963 (Tandy, 1978) have employed the services of landscape consultants to transform the appearance of their plantations (Campbell, 1977). Of course the results of these more enlightened practices will not be felt for several decades to come when the current plantations have matured into great forests similar to those of the Alps and Southern Germany, and have been diversified by sound management practices, including irregular rotations, so that one can walk between the trees rather than round the compartments (Goodall, 1973). In spite of this optimism, and the undoubted strides that have been made in reconciling land use competition and conflict between forestry and other uses, one still hesitates to endorse fully the most strident claims made for the role of forestry, as applied terrestrial ecology, by one Forestry Commission officer Holtam (1976, 123):

> Destruction of Britain's forests during 4,000 years had severe ecological effects, but plants and animals had time to adapt to a denuded landscape. In only 50 years foresters, through applied ecology and supported by sustained research have created on deserted, impoverished land, new forests to restore a part of Britain's plundered wealth. Plants and animals are adapting to the new forests which, in addition to growing wood at three times the rate of former forests in more fertile environments, provide our richest wild-life resource for centuries. Patience is necessary: the prospects for greater enrichment through continued research are great.

Mining

In the field of mining, however, it is unlikely that any mining operator would claim that mining did not compete and conflict with other land uses, both directly for the same site and indirectly by affecting adjacent land. However, they would normally claim that their operations can be made compatible with the environment, particularly by restoring land after mining has ceased. Nonetheless as Spooner (1981a) has pointed out, as societies become more affluent, demand for consumer goods slackens in favour of the 'superior' goods of amenity resources and environmental quality. Society therefore becomes less tolerant of the 'externalities' produced by mining operations, namely polluted land, air and water, spoliation of scenic beauty and increased noise and traffic. Unfortunately, as society becomes more affluent its consumption of mineral-based products and fossil fuels also increases, as shown in Table 3.2, and the future will in all probability be one where increasing annual tonnages of minerals will be extracted, with an increasing amount of land being affected. There will also be more emphasis on opencast and surface working, and a pattern of extraction in which fewer but larger pits, quarries and mines, and fewer but larger operators will have an increasing predominance (Blunden, 1975).

In common with forestry, one of the central issues in mining is the time lag between the decision to invest and the period when profitable extraction can start, as are the factors of world exhaustion of supplies, and the availability and price of supplies from overseas. In consequence, any decision to exploit minerals not only has to be a long-term one, but is also as in the case of forestry, a decision full of risk and uncertainty. Furthermore as the range of interaction between mining and other operations has become more complex, even in a vast and mineral-rich country like Canada (Marshall, 1982) the decision is no longer a straight economic or technological choice between competitive uses, but one which has to also take into account wider environmental implications and the criticisms of the environmental movement. However, unlike forestry which has managed to transform a bad environmental image into a relatively good one, mining as it has become bigger and more intrusive has become more of a land use issue.

The two main issues of interest to the rural geographer are first, the problems of derelict land once mining has ceased, and second, the huge spoil heaps left by the waste material, since many minerals are only found at very low levels of

Table 3.2: Production of minerals in the UK 1950–70

Mineral '000 tonnes	1950	1970	% change 1950–70
Iron ore	13,170	12,017	− 8.7
Chalk	13,137	16,063	+ 22.2
Clays and shales	23,962	31,307	+ 48.1
Sand and gravel	39,931	104,931	+ 162.7
Igneous rocks	11,423	36,958	+ 223.5
Limestone	25,363	87,808	+ 246.1
Salt	3,286	5,698	+ 73.4
Coal deep-mined	207,366	136,652	− 34.1
Coal opencast	12,090	7,925	− 34.4

Source: Blunden (1975).

purity. In some areas and for some minerals some of these problems have been solved by the same concepts of multiple use that have been developed in forestry. This is principally true of the group of minerals which have shown an about average increase in their use, as shown in Table 3.2, notably sand and gravel, and clays and shales. Since these are used in construction, and are primarily found around and used in towns, and are usually exploited in fairly shallow sites in river valleys, the sites can easily be flooded and turned into multi-purpose conservation and recreation sites, like the Cotswold Water Park (Collins, 1972).

However, the minerals that show the largest increase in Table 3.2, notably igneous rocks and limestones, are usually found in the far more sensitive areas, either the wild and more remote areas of upland Britain or the scarplands of southern England. In both cases these areas are also often designated as areas of landscape protection. Since the landscapes of these areas have already been dramatically transformed by either the afforestation or the agricultural revolution of the 1950s and 1960s, proposals to create even more visible and noisy intrusions have, not surprisingly, created severe conflicts of interest. In virtually every case however, the economic interest has triumphed, and in the last few years the mineral industry has had to compromise its objectives far less than the forestry industry. Prime examples of this are provided by extensions to china clay workings on Dartmoor, and to limestone workings in the Peak District National Park (Spooner, 1981a). In both cases, the argument that these are the only sites for the commercial exploitation of special deposits has been a crucial one.

Mining issues, however, are not confined to the urban fringe and the uplands, and first iron ore production (Gregory, 1971) and now coal production has begun to move away from the surface or shallow deposits around the upland areas of Britain, to often deeper but much richer deposits in hitherto unspoilt countryside (Spooner 1981b). Therefore the central mining issue in recent years has been the extension of coal mining to the farming vales of rural Yorkshire and the East Midlands. However, the first such development at Selby in Yorkshire was not strongly opposed in 1975, partly because of the oil crisis at the time. Indeed the inspector's report into the need for the mine recorded (Spooner, 1981b, 44) that 'he could find no outright objection to the grant of planning permission'. In contrast, the application to mine coal in the Vale of Belvoir in the East Midlands aroused a furore in the late 1970s and early 1980s

(North and Spooner, 1978). This resulted in a massive public inquiry which lasted for six months and cost £2 million (DOE, 1982a). After another delay of 15 months the Secretary of State for the Environment rejected planning permission for two of the three mines, but the third mine at Asfordby was given permission in 1982 after a revised plan had been submitted (Gilg, 1983b).

The cost and delays involved in this process were one reason why in the early 1980s the Commission on Energy and the Environment conducted a major review of the long-term environmental implications of coal supply beyond the end of the century. The review when published in 1981 (DOE, 1981, 205) attached 'The utmost importance to a continuing and forward-looking dialogue between developer, local authority and interest groups' as part of the search for a solution to the growing conflicts between mining and other land uses. However, the conflicts are likely to remain, since on the one hand, North and Spooner (1982) have commented that the report was more favourable to the coal mining industry than it was to the planners or environmentalists, and on the other, Manners (1981) has argued that Britain does not need both the coal and nuclear investment programmes envisaged for the end of the century. We can therefore conclude this section with the point that mining suffers from many of the same conflicts that forestry does, namely internal uncertainty about its investment programmes and markets in the medium to long term; external uncertainty regarding its position *vis-à-vis* other land users; and the extent to which other pressure groups and sectoral interests will use land use planning or other powers to circumscribe its activities.

Land use competition

Britain is a small and overcrowded island, and all the demands for the extensive use of space made by the land uses of agriculture, forestry and mining – already considered – and those also made by the land uses not so far examined – for example, water gathering, defence training, conservation and recreation – can at times combine together to produce almost intolerable conflicts and an apparently unresolvable competition between land uses. In the world at large, similar problems are increasingly becoming a feature of the rural geography of many countries, and even the USA, which at one time appeared to have limitless horizons, now faces severe problems of land use competition in some areas (Clawson, 1975b). Nonetheless, there are still very few countries, apart from Japan and Holland, which experience the degree of competition found in Britain, and in particular, the intense competition found in the British urban fringe. Indeed Munton (1983) has described the conflicts between the interests of agriculture and all other land uses in the urban fringe as endemic.

Because agriculture is the main land user in the urban fringe and the one under most threat, most studies have examined either or both the rate of loss of farmland and the problems facing agriculture. For example, Low (1973) found that the West London borough of Hillingdon lost 530 of its 2,500 hectares of farmland between 1961 and 1971 and that all the farmland could disappear by the year 2001. However, not all 'open land' would disappear since farmland was being lost not only to urban growth, but also to recreational and mineral interests. Nonetheless, large areas have been and will continue to be lost. This is not only a problem by itself for an island which is only 62 per cent self-sufficient in food (MAFF, 1983a) but also because urban fringe land is often some of the best and most productive farmland. For example, in the USA the areas adjacent to the main metropolitan areas contain nearly 52 per cent of the prime land (Coughlin, 1980), and if current rates of loss are projected to the year 2001 it has been estimated that between 2.8 and 5.5 per cent of all prime land could be lost between 1975 and the end of the century.

However even where, as in Britain, planning controls have effectively halted urban growth into many urban fringe areas (Munton, 1983), particularly those with green belts, the farms that remain still face considerable problems from other sources. For example, Thomson (1981) in a study of the area around London and the six metropolitan counties of England, found that farms in the fringe areas (about 10 miles from the town centre of the metropolitan areas and 20 miles from the centre of London) had low-input, low productivity farming which had been unable to expand or develop as far as farming elsewhere. Apart from the obvious disincentive to invest in better farming where the farm may be taken for urban development or a more profitable use like a golf course, Thomson also found that the main threats to the farms were fragmentation by roads, houses and factories; shortage of land for expansion; sale of land for development; and trespass and vandalism. The picture was not entirely one of gloom however, and Thomson not only found pockets of intensive horticulture and dairying, but also pointed out the advantages to farmers of being close to a big population centre for buying and selling goods, and employing full or part-time labour. Indeed, some of these advantages must have been in the minds of the 72 per cent of farmers in one survey (Rettig, 1976) who said that they would continue to farm in the urban fringe of Sunderland even if they had the opportunity to farm elsewhere. However, this figure could be regarded as rather low in the light of all the evidence that shows farmers to be relatively immobile (Gasson, 1969).

It also reflects the fact that most of the studies of urban fringe agriculture have been rather qualitative with regard to the actual effects of the urban fringe, in that they have listed the numerical occurrence of various problems but not quantitatively evaluated their effect. In an attempt to remedy this situation, White and Silverwood (1983) have proposed three quantitative techniques for: (a) the evaluation of the costs and benefits of the urban fringe on a field by field basis; (b) an assessment of the effects of the urban fringe on overall farm viability; and (c) the development of a predictive computer model.

Although this and similar work will no doubt enhance our understanding of the problems of land use in the urban fringe, and to some extent the problems of land use competition elsewhere, the greatest needs are: (a) for more satisfactory methods for equating the arguments for and against one particular land use with another, and (b) for more satisfactory methods of reconciling land use conflicts, whether by positive land management, or by negative land use control (Munton, 1982) or by a mixture of the two. Willis (1982) has proposed a formula for evaluating the net social benefit (NSB) of green belt policy. Under this formula the net social benefit comprises the difference between the value of the factors of production and consumption under a green belt policy and their social opportunity cost, and is calculated as:

$$NSB = \sum_{t=o}^{T} \frac{\sum_{j=1}^{n} Qtj \, (Ptj\text{-}Ptj^*)}{(1 + i)^t}$$

where Qt   = Quantity of factor j used
       Pj   = Value of factor or good j under green belt policy
       Pj* = Social opportunity cost of factor or good j
       T   = time period of the policy, with t denoting specific years from o to T
       i   = discount or time preference rate

In recent years both these needs have captured the attention of more and more rural geographers as the subject has moved away from description alone to not only explanation, but also via methodological and technical developments to problem solving and policy formulation. These new concerns and interests are more fully considered in Chapters 8 and 9 which focus on studies of landscape and land use change, land management and planning and rural development.

Before these topics can be considered, however, other uses of rural land both intensive and extensive, as well as the social and economic fabric of rural areas, must be examined. Accordingly, the next few chapters turn away from the more traditional fields of rural study – the farm, the forest and the mine – to the newer areas of concern: rural settlement and housing, rural population and employment, rural transport, service provision and deprivation, and rural recreation and tourism.

4

Rural settlement and housing

Rural geographers have been interested in rural settlement and housing for a very long time. For example, some of the earliest studies and theories related rural settlement to geology and the landscape (Penoyre and Penoyre, 1978), and produced relationships like nucleated villages in arable areas and dispersed villages in pastoral regions (Roberts, 1977). However, in the past 30 years these ideas have been dismissed as too deterministic and merely descriptive, and attention has shifted instead towards the development of theoretical models of settlement patterns, and then more recently to the implementation of these models in rural planning policies. At the same time increasing attention has been paid to the individual components of rural settlements, the houses themselves, and perhaps, in a reaction to the theoretical-economic dominance of recent years, some work on the relationship of settlement to landscape has begun to appear again, albeit mainly in the historical rather than the contemporary field (Rowley, 1978). Accordingly, this chapter is divided into the following sections: *theoretical approaches to rural settlement*; *rural settlement policies*; and *rural housing*.

Theoretical approaches to rural settlement

The first issue that any theory about rural settlement has to confront is the definition of rural settlement to be employed. This can vary widely both between and within countries (Bunce, 1982) and also within the sub-classifications of hamlet, village and small rural town, but for the sake of argument this chapter initially follows the definition of a village employed by Moss (1978) namely, a population between 200 and 5,000 people. Extending this definition, it also seems fairly wise to define hamlets as those settlements below 200 people, and small rural towns as those with populations between 5,000 and 10,000. However in the USA, Johansen and Fuguitt (1984) have used an upper limit of 2,500 people for villages.

Most theories of rural settlement have taken basic classifications of this type and attempted to relate them to certain variables. For example, Clark (1975) in a study of 44 rural districts between 1563 and 1911 was able to show that 'population attracts population' and that the proportion of larger villages had increased at the expense of the proportion of smaller villages as shown in Table 4.1.

Clark's empirical work confirmed Chisholm's (1962, 172) inescapable conclusion based on location theory and empirical evidence that 'At all scales, in the economically more advanced nations, powerful economic forces are working for an increasing degree of agglomeration.'

Table 4.1:　Change in village size 1563–1911

Size	Number of villages			Ratio 1911/1563
	1563	1801	1911	
0	298	298	286	0.96
50	266	293	284	1.07
100	197	271	268	1.36
150	127	243	238	1.87
200	97	212	220	2.27
250	66	148	173	2.62
375	34	91	103	3.03
500	25	57	72	2.88
1,000	7	36	52	}
2,000		17	19	} 11.71
3,000		11	11	}

Source: Clark (1975).

It is these economic forces that have been the main explanatory variables used in most of the theoretical models used to explain the location, size and spacing of settlement patterns. Without a doubt the most famous of these are the so-called 'central place' models developed by Christaller (1966) in the late 1930s and Lösch (1954) in the immediate postwar years. These are too well known to need a thorough description here, but since they have provided most of the concepts which have been employed by contemporary rural geographers and rural planners seeking to adopt them in actual policies, the following basic principles should be recalled.

First, the models assume that the major function of rural settlement is to act as a market or service centre. Therefore the most important variable is not the size of a population but its centrality which can be measured by two concepts:

1　The 'Threshold Population' which is the minimum population necessary for a good or service to be provided profitably, and

2　The 'Range of a Good' which is the maximum distance people will travel to purchase a good or service.

It is clear from these two concepts that each Central Place would serve a circular trade area, but because overlapping trade circles would lead to an inefficient service, the models use hexagons instead to define the market area of each Central Place.

However the 'Threshold Population' and the 'Range of Good' varies for each good or service, and although this variation in practice is more of a continuum than a grouped hierarchy, Christaller, from empirical observation in Southern Germany was able to develop the theoretical idea of a settlement hierarchy, as shown in Table 4.2 and Figure 4.1, in which settlements and their hexagonal market areas would be nested within each other at regular intervals.

Christaller also produced a number of variations based not only on the market principle, but also the 'Traffic Principle' where transport costs are high, and the 'Administrative Principle' for administrative convenience. In a later development of the model, Lösch (1954) returned to the idea that each good has its own threshold and market area, and produced 150 overlapping networks for each market area, and then superimposed and reoriented them, to minimize the number of and distance between central places.

Table 4.2: Characteristics of central places in Southern Germany

	Number of places	Distance apart (km)	Area of region (sq. km)	Number of goods offered	Population of place	Population of region
Marktort	486	7	44	40	1,000	3,500
Amtsort	162	12	133	90	2,000	11,000
Kreisstadt	54	21	400	180	4,000	35,000
Bezirksstadt	18	36	1,200	330	10,000	100,000
Gaustadt	6	62	3,600	600	30,000	350,000
Provinzhauptstadt	2	108	10,800	1,600	100,000	1,000,000
Landeshauptstadt	1	186	32,400	2,000	500,000	3,500,000

Source: Christaller (1966).

Although both Christaller's and Lösch's models can be criticized as being over-economic and over-dependent on the concept of an isotropic plain, they have given rise to a number of related models and empirical settlement hierarchies, and have also been widely employed in settlement planning, as the next section demonstrates. One major theme common to both developments has been the idea of thresholds, and the division of settlements into sizes, depending on the range of goods and services they can support. For example, Edwards (1971) in a study of northeast England divided the area's settlements into eight grades as follows:

Grade 1 Centre: Regional capital
Grade 2 Centres: 5 settlements: 4,500 to 12,500 adults
Grade 3 Centres: 23 Large market settlements: 900 to 4,500 adults
Grade 4 Centres: 33 medium market settlements: 411 to 900 adults
Grade 5 Centres: 56 agricultural service settlements: 165 to 410 adults
Grade 6 Centres: 58 small villages: 90 to 164 adults
Grade 7 Centres: 130 hamlets: 36 to 89 adults
Grade 8 Centres: Small hamlets and farmsteads: less than 36 adults

Edwards concluded from a study of population change in the eight grades that some were far more susceptible to population decline, notably grades 5, 7 and 8, and that grades 4 and 6 seemed to mark bottom-line thresholds for settlement survival.

Other workers approached the issue from different directions and instead of using market centres (Christaller) or population change (Clark and Edwards) they looked at the cost of public service provision. For example, Warford (1969) in a study of rural Shropshire worked out the costs and benefits of relocating the population into fewer but larger settlements, to minimize the cost of supplying water. He also examined the savings there would be in providing other public services like education and postal services for various discount rates and concluded, as shown in Table 4.3, that the savings to be made out of relocation and concentration could be impressively high even in 1969 money.

Although cost/benefit analyses of this kind can be criticized as being unrealistic and subject to a wide margin of error, another study along similar lines in northeast Norfolk (Norfolk County Council, 1976) confirmed the savings that might be achieved by concentration. For example, overall annual revenue costs could be about one-third lower under a policy of concentration, compared to a policy of continued dispersal (Shaw, 1976) and this type of

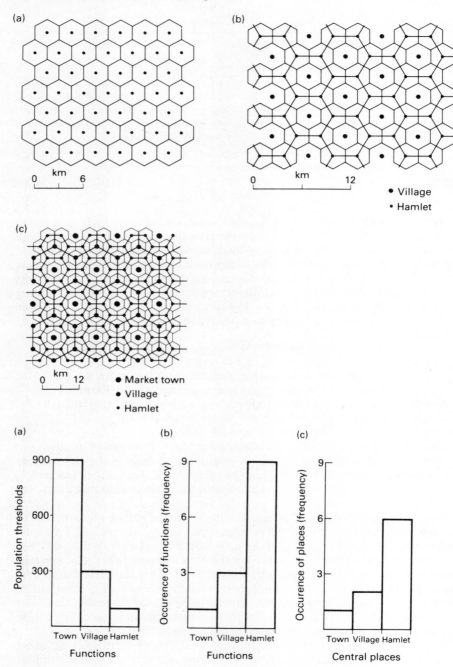

Fig 4.1 Development of central place theory and concept of thresholds. The top diagram shows the build-up of nested hierarchies of central places from (a) hamlets to (c) market towns with (b) villages in between. The lower diagram shows the concept of thresholds and the number of functions on which the theory is based. Source: Chapman (1979)

Table 4.3: Savings to be obtained from relocation depending on rate of discount

	Four rates of discount (£ thousands)							
	A	B	A	B	A	B	A	B
	6%		8%		10%		12%	
Water supply	852	508	825	490	806	477	792	468
Sewage disposal	352	283	335	272	322	264	312	259
Telephones	16	5	13	4	11	4	9	4
Electricity	39	27	33	26	29	25	26	25
Mail	156	95	129	78	110	67	95	58
Schools	373	322	314	278	271	245	238	221

Note: A = No relocation of population.
　　　　B = Total relocation.
　　　　A − B represents savings to be made by relocation.
　　　　So at a 6% rate of discount the saving on telephones would be
　　　　£16,000–£5,000 = £11,000.
Source: Warford (1969).

finding has encouraged the development of models of the theoretical rationalization of settlements shown in Figure 4.2.

Both the Shropshire and North Norfolk studies are broad examples of a type of theoretical approach that came to be known as threshold analysis in the early 1970s (Kozlowski and Hughes, 1972). Under this theory three groups of costs are applied to running or expanding a settlement. First, the costs imposed by physical constraints, second, the costs imposed by existing infrastructural capacities, and third, the costs of providing new infrastructure when population growth forces the development of new capacity (SDD, 1972). The theory argues that settlements should therefore be classified into stepped thresholds relating to the cost of providing each new grouped set of infrastructure. However, although the theory has been seen to operate in practice, for example in Ivybridge (Glyn-Jones, 1977), where the provision of one new service led to the development of other new services in a domino-type pattern, Simpson (1977) has argued that all thresholds above the bottom line are in practice very difficult to define, and that in reality, a continuum of thresholds for different services exists, in parallel with the continuum of thresholds for market services that also undermines central place theory.

Nonetheless, both central place and threshold theory seem to indicate that settlements naturally fall into certain groups and that certain minimum sizes are necessary for survival. Therefore Pacione (1982) in his study of rural settlements in the Strathclyde region of Scotland tried to test the viability of smaller rural settlements, since neither central place theory or the classifications produced by Dickenson (1947) and Bracey (1956) provided a comprehensive assessment of settlement viability. Accordingly he used a multivariate statistical approach based on 46 census variables, and after subjecting these to an R-type principal components analysis, he concluded that there was a clear preference for the more rigorous taxonomic procedure of cluster analysis. This produced a five-cluster grouping and Pacione argued that this not only simplified the task of identifying less viable settlements, but also provided decision-makers with valuable guidance for the allocation of scarce resources.

Indeed the evolution of rural settlements is now strongly influenced by the

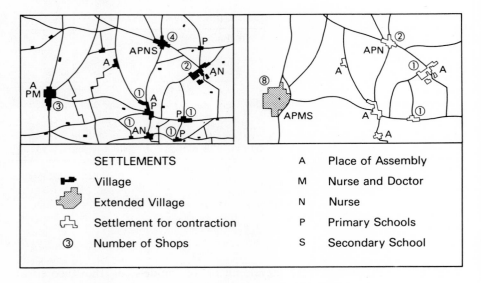

SETTLEMENTS

		A	Place of Assembly
	Village	M	Nurse and Doctor
	Extended Village	N	Nurse
	Settlement for contraction	P	Primary Schools
③	Number of Shops	S	Secondary School

Fig 4.2 Theoretical rationalization of the settlement pattern. The diagram on the left shows an existing pattern. The diagram on the right shows a possible pattern based on the reduction of servicing costs. Source: Gilg (1979)

decisions taken by land use planners, and those in charge of the provision of public services and Lonsdale (1981) has argued that this is in fact the critical issue. Therefore, theoretical approaches like Hudson's (1969) three-phase biological process of colonization, spread and competition are not only restricted to their own specific cultural and technological context, in this case Iowa, but are also not found to operate universally (Grossman, 1971). Nonetheless Haining (1982) has developed Hudson's model by introducing a scale-dependent statistic that measures second-order point patterns on maps. But Grossman (1983) has pointed out that the model totally omits man-made infrastructure, a major problem since all rural settlements depend on such infrastructure. A further problem, as Hudson himself admits, is that such theories also omit the role of the government in shaping rural settlement patterns. Therefore, attention is now directed towards the role of the state in producing rural settlement policies, and in particular, to the way in which these policies have employed the theoretical approaches already discussed.

Rural settlement policies

In his authoritative *Introduction to Rural Settlement Planning* Cloke (1983, 72) argues that:

> It is clear that definite links can be drawn between various formative theories and the course taken by post-war rural settlement planning.

However, Cloke also argues that two factors temper the importance of these links; first, the theories may have been used to justify policies founded on economic and administrative expediency, and second, the actual decisions that are subsequently taken often bear little relation to either the policy or its assumed theoretical justification. Nonetheless Cloke believes that four theories have been directly influential in the production of rural settlement plans, namely:

1   A hierarchical settlement pattern (Central Place Theory);
2   Service Thresholds;
3   Economies of Scale; and
4   Growth centres.

All of these tend to lead to policies of centralization and in particular to the idea of 'key settlements' in which both population growth and service provision are concentrated. Woodruffe (1976) argues that the 'key settlement' concept has been extensively used in many parts of Britain, and Cloke (1983) has also shown how these policies have been used in both periods of plan making, the Development Plans of the 1950s, 1960s and early 1970s and, albeit in a more modified form, in the Structure Plans of the late 1970s and early 1980s. Nonetheless Cloke also points out that within the Development Plan period that not one but three main types of policy can be identified:

1   *Key Settlement policies*, as shown in Figure 4.3, in which a comprehensive policy of concentration of housing, employment, services and facilities into selected centres is followed, in order to maintain a viable level of rural investment, to support not only the key settlement but also its hinterland villages;
2   *Planned decline policies* where a direct rationalization of the settlement pattern is pursued; and
3   *Village classification policies* where settlements are categorized by environmental quality and service capacity, so that optimum use may be made of existing public investment, by encouraging new development into such, usually larger, villages.

Within these Development Plan policies the more rural counties with declining populations often employed a generous key settlement policy, but the 'pressured' counties around the major conurbations often used a very tight key settlement policy with only a few settlements chosen for growth.

When the Development Plans were replaced by Structure Plans in the mid-1970s to early 1980s (for a commentary see Chapter 9) some reappraisal of their dominant key settlement ethic had taken place, and Cloke and Shaw (1983) in a survey of 35 county planning authorities found that four problem areas (Cloke, 1983) had been identified with the key settlement concept. First, the assumption that the hinterland of key settlements would benefit from their improved facilities had been undermined by a decline in accessibility, second, the policy had aided and abetted the process of rationalizing and cutting down on rural services, third, the level of investment in all rural areas had been small and so the effects of concentration had been minimal even in the key villages, and fourth, land use planners in practice had had little power over the provision of public services and these had often been located in non key settlement villages.

Not surprisingly the Structure Plans have been less dominated by the key settlement concept and five other types of policy have been recognized by Cloke and Shaw (1983) in their survey of virtually all the English and Welsh Structure Plans. As Figure 4.4. shows, these are variations rather than radical departures from the key settlement concept, and key settlement type policies are still found in 16 plans, and similar policies are found in the seven plans with 'tiered' (but

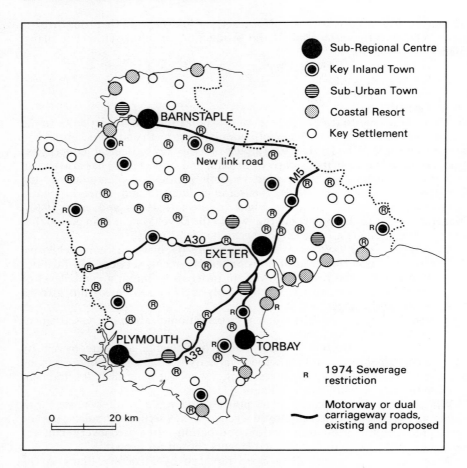

Fig 4.3 Key settlement plan for Devon. The diagram also shows that a large number of key settlements could not in fact grow in the short term because of a sewerage restriction. Source: Gilg (1979)

not key settlement policies) and the eight plans with 'hierarchical restraint' policies.

More radical variations are found in the three plans which have adopted a 'concentration in market towns' policy, the ten plans with 'severe restraint policies', and the five plans with 'area approach policies'. The 'market towns' policy is related to the argument that small rural towns that have otherwise lost their economic *raison d'être* (Coates, 1977) can and should be expanded by taking overspill from congested urban areas. However Moseley (1973a) has found that there can be problems with merging employment and amenities in this approach, and that 'trickle-down' effects to the wider hinterland of the town can be minimal. Nonetheless, Moseley (1973b; 1973c) has still concluded that the towns themselves and also their immediate area, in both Brittany and East Anglia, have benefited from such schemes.

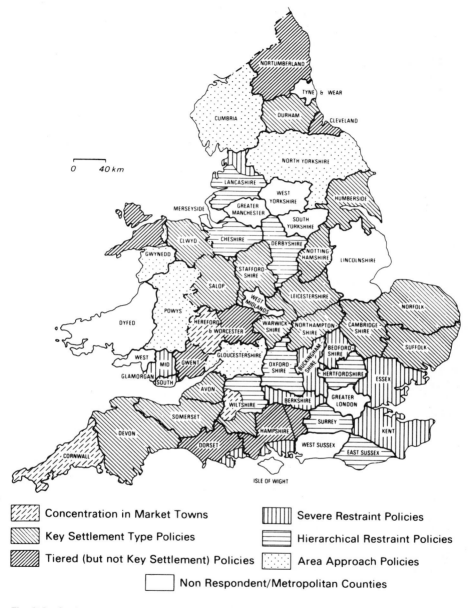

Fig 4.4 Settlement policies employed by County Structure Plans. Source: Cloke and Shaw (1983)

Expanded towns beyond the urban fringe or green belts are a related policy to the ten plans employing 'severe restraint' policies, which as Figure 4.4 shows, are all found around London, Southampton, Cardiff and Lancaster. But in this case the aim is also different, in that the overall objective is the protection of the countryside from urban sprawl, rather than the planning of the settlement pattern.

It is too soon to say what the effects of these more diverse rural settlement policies will be, but a number of studies and critiques have been made of the previous Development Plan period which was dominated by the key settlement concept. These studies can be divided into two types, first, studies of the impact of the policies, and second, critiques of the key settlement concept itself.

The first group of studies into the direct impacts of the policies are only briefly reviewed here since they are reviewed more fully in Chapter 9, but since they have a bearing on the more theoretical critiques of key settlement policy, an outline at this juncture would be useful. Cloke (1979) in his study of two rural counties, Devon, a 'remote' rural area, and Warwickshire, a 'pressured' rural area, concluded that the policy hadn't been too successful in achieving its twin aims of (a) concentrating population, employment and infrastructure and (b) improving or stabilizing opportunities for the residents of hinterland settlements, for three main reasons. First, the wrong key settlements may have been chosen, second, development control decisions didn't always follow the policy, and third, rural land use planners don't have sufficiently wide powers to provide effective answers to the problems of transport, housing, employment and services.

Another reason advanced by Cloke, the reliance on restrictive rather than positive planning, was confirmed by Martin and Voorhees (1981) in their study of six district councils. They concluded that key settlement policy is more able to halt development than to initiate it, and is thus more suitable for pressured areas, rather than areas of slow or declining population change. They also argued: (a) that key settlement policy had only achieved modest success in respect of social and economic objectives; (b) that it has had only a minor impact on the long-term historical development of settlements; and (c) that there was no conclusive evidence that concentration policies had achieved the most economical pattern of infrastructure and service provision.

More localized studies have tended to confirm these findings. For example, Blacksell and Gilg (1981) in their study of Devon found that development has been confined to existing settlements but not necessarily the key settlements, and Phillips and Williams (1983a, 512) in a study of South Devon, argued that there had been few attempts to harmonize the work of the planning and housing departments and concluded:

> The major implication is that the key settlement/selected local centre level of the settlement hierarchy is largely redundant in terms of the practical application of local authority housing,

However, Anderson (1981a) in a study of East Sussex concluded that the strategic policies of the Structure Plan (according to Cloke (1983) an example of a 'Hierarchical Restraint Policy') were being effectively implemented through development control decisions, and also argued that this could partly be due to a long history of restraint. This is a point confirmed by Cloke (1979, 226) who argues that key settlement policies have only had 25 years to prove themselves and should thus be given more time, since:

Clearly, some physical, economic and even partial social success has already been achieved through the use of key settlement policies.

However, a second group of studies into key settlement policy have been more critical, especially in their theoretical critique of the underlying assumptions of the policy. Cloke (1983) has classified these critiques into four groups:

1 Key settlements have no theoretical basis, and key settlement theory has often been invoked in a post-hoc rationalization exercise, to justify strategies of resource concentration and public expenditure cuts. Furthermore, little empirical evidence has been found to justify the arguments that key settlements in practice act as growth poles (Grafton, 1984).

2 Key settlements exacerbate the social problems encountered by many rural residents, and as the section on 'Rural Deprivation' shows in Chapter 6, key settlement policies have failed to meet the housing, employment and transport needs of those living outside the key settlements, particularly the young, the old and the housewives (Hanrahan and Cloke, 1983).

3 Key settlements are politically unworkable, and although Anderson (1981a) has shown a good correspondence between policy and practice other workers, Blacksell and Gilg (1981) and Cloke (1979) have shown a lack of correspondence between the two. In the 1980s this position could only become worse since the rural settlement policies, shown in Figure 4.4, are produced by the county councils in their Structure Plans, but are put into practice by district councils, who only have to have regard to the provisions of the structure plan, and can in effect, make their own policy to a large extent.

4 Key settlement policies are not the most cost-effective mechanism for rural resource allocation, and the so-called economies of scale they are supposed to produce do not in fact exist.

The most effective critique along these lines has been provided by Gilder (1979) in his study of the area around Bury St Edmunds in Suffolk. He found, as shown in Table 4.4 not only that the population had grown but that the growth had been largest in the villages with a population above 1,000. However, as Figure 4.5b shows, when actual unit costs rather than costs per head are computed, the economies of scale shown in Figure 4.5a cease to exist, not only for education but for all services as shown in Figure 4.5c. Gilder suggests

Table 4.4: Population change in the Bury St Edmunds area 1961–76

	1961			1976		
	Number of settlements	Population	%	Number of settlements	Population	%
Bury St Edmunds	1	21,179	43.1	1	29,200	44.2
Villages above 1,000	4	4,502	9.2	10	15,500	23.5
500–999	13	10,813	22.0	14	10,320	15.6
250–499	22	7,407	15.1	16	5,870	9.0
Under 250	37	5,295	10.6	36	5,060	7.7
Totals	77	49,196	100	77	66,050	100

Source: (Gilder, 1979).

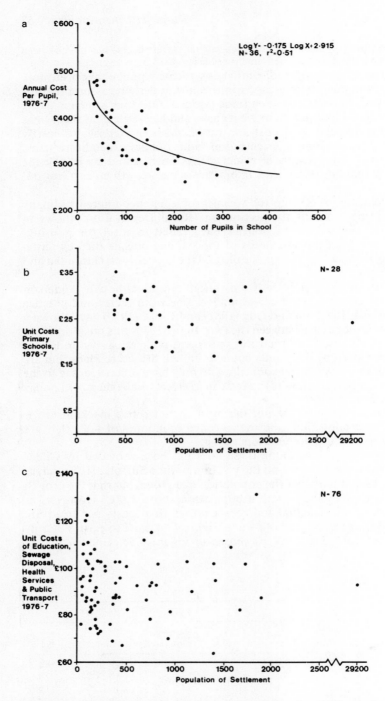

Fig 4.5 Unit costs and costs per head of providing rural services. Unit costs do not show the same economies of scale as costs per pupil. Source: Gilder (1979) Reproduced with the permission of Pergamon Press. (a) Top diagram (b) Middle diagram (c) Bottom diagram.

therefore that economies of scale are insignificant in villages having a population of up to 2,000 and that the unit costs in these villages are not significantly higher than in a town of 29,000 (Bury St Edmunds). Gilder argues that the expected economies of scale are overwhelmed by variations in standards of service, demographic factors and potential changes in the operating methods of rural services. Although these conclusions have been queried, and Gilder has accepted that the methodology needs refinement (Curry, 1981), Gilder's two suggestions for the future are probably still broadly valid. First, that any attempts to minimize public service costs may best be achieved by an empirical rather than a theoretical investigation of costs, and second, that in the study area at least (Gilder, 1979, 264) that:

> The accommodation of future growth will be less costly if that growth is dispersed widely throughout the area, than if it is concentrated in Bury St Edmunds and the large villages. The marginal costs of making better use of existing schools and other fixed assets is likely to outweigh the benefits of concentrated development policies, even if economies of scale within services exist.

Arguing along the same lines McLaughlin (1976) has put forward a theoretical case for a set of functionally interdependent villages. Under this proposal, the planning unit would not be an individual village but a system of perhaps five to six linked villages over which public services, education, residential and employment opportunities would be spread. Cloke (1977a) in a response, however, argued that it was too soon to reject key settlement policies, and that too much investment had already been made in this for the policy to be reversed. However, six years later Cloke (1983) had modified his position and argued that key settlement policies should be adapted: (a) to place more emphasis on socioeconomic planning and the elimination of deprivation; and (b) to place less emphasis on physical planning and conservation. But he still acknowledged (Cloke, 1982a, 284) that much of the debate still centred around the concentration/dispersal debate although 'a redirection towards resource dispersal seems unlikely', and this was confirmed by a survey of county planning authorities (Cloke, 1982b) which showed that 20 authorities didn't think further resource dispersal was desirable compared to only 12 who thought it was.

Nonetheless, Cloke (1982a) has also outlined four future courses of action, ranging from a 'do nothing' approach to a policy of positive discrimination towards the most deprived rural communities. Other radical proposals have been put forward by the Rural Resettlement Group (1979) who have provided a guide for the considerable number of people, many of them professional groups, who wish to relocate their lives and employment in the countryside (this repopulation of the countryside is discussed more fully in Chapter 5), and by M. Anderson (1975) and Boddington (1984) who have argued that increasing agricultural self-sufficiency could release up to 3.5 million hectares of farmland, and that this land could be taken over by a massive and widespread rural building programme based on 20-hectare smallholdings, thus smashing wide open traditional policies of village containment based on key settlement policies.

More realistically, the pressure for new villages continues to grow, particularly in central and southern England, and although only one has so far been built, New Ash Green in Kent with a planned population between 6,000 and 7,000 (Bray, 1981). A number of expanded villages like South Woodham

Table 4.5: Some indices of rural housing

A.   Ratio of sub-standard dwellings in each type of area to the average proportion in England as a whole 1981.

	Conurbation	Provincial town	Rural
Unfit dwellings	1.05	0.86	1.10
Dwellings lacking one or more amenities	1.14	0.92	0.84
Repairs cost over £7,000	0.91	0.79	1.49

B.   Housing tenure in English rural districts early 1980s.

Tenure %	Rural districts	England as a whole
Owner-occupied	60.6	57.6
Local Authority	21.6	28.8
Other rented	17.7	13.6

C.   Rural housing and rural incomes in two English districts 1977.

Annual income	Owner-occupiers	Council houses tenants	Private tenants*	Tied housing	Houses lacking amenities	Houses lacking a car
South Oxfordshire %						
Less than £3,000	22.6	50.0	68.6	31.5	68.4	66.0
£3,000–£6,000	37.3	39.3	22.9	48.1	31.6	30.9
Over £6,000	40.1	10.7	8.5	20.4	0.0	3.1
Cotswold District %						
Less than £3,000	35.2	63.9	59.4	60.3	76.9	84.0
£3,000–£6,000	32.4	33.3	31.3	33.7	23.1	16.0
Over £6,000	32.4	2.8	9.3	6.0	0.0	0.0

* Unfurnished.
Sources: Phillips and Williams (1984); Clark D. (1984) and Rogers (1983).

Ferrers in Essex contain many of the same characteristics, and in 1983 a consortium announced plans to build a series of 12 new villages in southeast England each with a population of between 15,000 and 20,000 (Gilg, 1984a; Merrett, 1984). These forces of expansion are not confined to the UK, and as Chapter 5 shows, the repopulation or counterurbanization of the countryside, particularly in small towns and large villages, can also be found in Switzerland (Wilkes, 1983) and the USA where Carpenter (1977), in a survey of potential migrants, found that the majority would prefer to live in settlements below 50,000 people but that any precondition, like a reduction of income, markedly reduced the willingness to migrate.

One of the main reasons for migration both to and from the countryside is the need for a house, particularly at the time of marriage (or in these days the first setting up of home), and at retirement, and so attention is now turned to a topic that has aroused increasing interest among rural geographers in recent years, rural housing.

Rural housing

Traditionally rural geographers were mainly concerned with the architecture of

Table 4.6: Pahl's typology of social groups relevant to rural housing

1 Large property owners: owner-occupiers who are also property landlords.
2 Salaried immigrants with some capital: owner-occupiers whose main influences are in the improvement and gentrification of rural housing.
3 Spiralists: transient owner-occupiers characteristically occupying a recent housing development in an accessible village.
4 Those with limited incomes and little capital: reluctant commuters who have been forced out of towns by the need for cheaper but larger housing in response to a growing family.
5 The retired: either as owner-occupiers or tenants in large villages where accessibility is high and service provision comparatively good.
6 Council house tenants: marks the transition from those who can buy and those who must rent housing, characterized by relative inflexibility of tenure and location compared with owner-occupiers.
7 Tied cottages and other tenants: the rural poor with low wages, poor housing and isolation.
8 Local tradesmen and owners of small businesses: rather indeterminate in its housing characteristics and may include both owner-occupiers and tenants.

Sources: Pahl (1966a); Pacione (1984).

rural housing and how their vernacular styles related to the landscape and geology of the area, but in recent years concern has shifted towards two very different themes: (a) the quality of the housing stock; and (b) differing opportunities of access to the stock particularly between various social groups. It is often wrongly assumed that slums do not exist in the countryside and the rose-covered cottage myth is partly to blame for this false assumption. However, poor quality housing in rural areas has long been a problem, and although attempts to deal with it date from the 1920s, and in particular from the 1940s (MOH, 1944) when 11.6 per cent of rural houses were found to be unfit and 33.4 per cent in need of repair (MOH, 1948), the situation was still unsatisfactory in the 1970s and 1980s. For example, the 1976 and 1981 House Condition Surveys found that though the percentage of unfit rural houses had fallen to just over 5 per cent in 1976, it had risen again to 6.7 per cent in 1981 (DOE, 1978a, 1982b), and moreover, as Table 4.5a shows, rural areas were worse off than urban areas, thus exploding the myth of better rural housing.

Table 4.5b also shows that rural dwellers are more likely to own or privately rent their homes than their urban counterparts but less likely to be council tenants. However, Table 4.5c shows that in two rural districts, owner-occupants are more likely to be found in the upper income brackets (note inflation since 1977), but that the more disadvantaged sections of rural society as measured by rented or tied housing, and housing lacking basic amenities and cars are found in the lower income bracket. Overall Table 4.5 shows that social class and tenure groupings are important variables in the rural housing field and indeed, since Pahl (1966a) first provided the typology of rural housing shown in Table 4.6 based on social class and tenure groupings, most workers in this field have followed this type of approach, for example, Phillips and Williams (1983b) and most notably Dunn, Rawson and Rogers (1981).

The study by Dunn *et al.* (1981) is in fact the most comprehensive survey of rural housing yet undertaken, and provides not only a wealth of background information but also detailed information about housing in two rural districts, the Cotswolds and South Oxfordshire. Its most important contribution, however, may have been the production of the seven rural housing profiles shown in Figure 4.6 and Table 4.7. These were produced from a cluster analysis of Small Area Statistics from the 1971 population census, which used 30

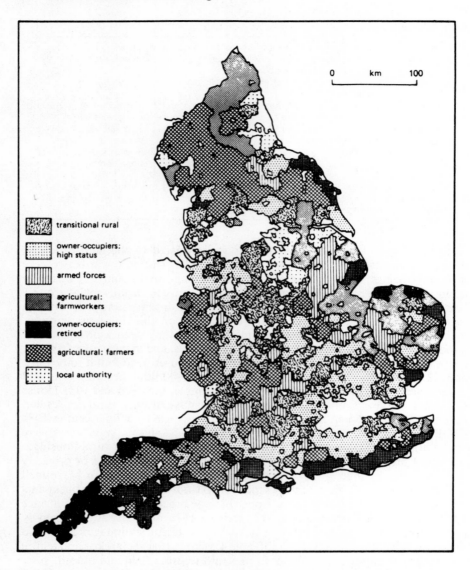

Fig 4.6   Rural housing profiles for England and Wales. For key see Table 4.7. Source: Dunn *et al.* (1981)

Table 4.7: The character of the rural housing profiles shown in Figure 4.6

Profile 1: Agriculture: farmworkers (42 cases)
   A high proportion of agricultural workers and low proportions of professional and non-manual workers. Above average private rented unfurnished tenancy group and also greater than average number of households lacking exclusive use of one or more basic amenities. High unemployment rates. Low level of car-ownership.

Profile 2: Agricultural: farmers (91 cases)
   A predominantly agricultural group with low population totals and high proportions of farmers and agricultural workers. Many households living in rented unfurnished accommodation. Low occupancy rates. Above-average proportion of pensioners.

Profile 3: Owner-occupiers — retired (42 cases)
   Large retired population. Low economic activity rates, high unemployment rates. High level of owner-occupation and under-occupancy of dwellings.

Profile 4: Transitional rural (90 cases)
   A large profile incorporating a variety of cluster characteristics. High female economic activity rate. Above average proportion of non-manual workers, slightly more than average percentage of children. Fewer pensioners and agricultural workers than the average for rural districts. Renting privately unfurnished is a smaller than average tenure category and there is less under-occupancy.

Profile 5: Owner-occupiers: high status (83 cases)
   High proportion of heads of households employed in professional and non-manual occupations and correspondingly above average level of education. Two-car ownership and owner-occupation are above average.

Profile 6: Armed forces (36 cases)
   A small profile dominated by armed forces personnel and households living in rented unfurnished accommodation. Above-average proportion of children leading to some overcrowding. High male economic activity rate. Few pensioners.

Profile 7: Local Authority housing (26 cases)
   A small group of rural districts with high proportions of households living in Local Authority rented accommodation. Skilled manual workers dominate the occupational structure. Relatively few households own cars. Some overcrowding of dwellings.

Source: Dunn *et al.* (1981).

variables relating to housing and household characteristics which were then aggregated from enumeration districts into the 410 rural districts which existed in 1971. The resulting map not only highlights tenure divisions but also the areas affected by commuting ((4) 'transitional rural' and (5) 'owner-occupier:high status') and the more rural areas less affected by urban influences, at least in 1971, for as Chapter 5 shows, there has been a marked repopulation of the more traditional rural areas since 1971.

   Another major contribution of the study was a survey of movement through the housing stock and competition between the different social groups, and although the authors argue (Dunn *et al.*, 1981, 246) that while 'owner-occupation may be the utopia for many, its attractions are not universal', they also highlight a growing discrepancy in rural areas, between a relatively prosperous owner-occupier sector, and an increasingly disadvantaged tenanted sector. This division has been growing because affluent urban immigrants can outbid local residents on poor rural wages, and also because key settlement policies and conservation policies in National Parks and Areas of Outstanding Natural Beauty have either limited the stock of new housing, thus inflating prices, or insisted on high standards of development which has led to an imbalance in favour of better quality and therefore more expensive housing. It is not surprising, therefore, that most recent work has focused attention on the special problems of the disadvantaged poor in the designated conservation areas, notably in National Parks, and in the two main rented tenure types, council houses and tied cottages.

In the National Parks, much attention has been given to the competition for the limited housing stock and Rogers (1981a, 194) has concluded that:

> Relatively poor local economies, few local authority houses and high house prices combined, encourage younger people to leave while incomers have been quick to buy houses taken out of the privately rented sector. The result is an increasingly old population and a housing environment which is poor for local people and increasingly open to invasion from outsiders.

One National Park Authority, the Lake District Special Planning Board, has tried two different solutions to this problem. The first policy, between 1956 and 1975, was unashamedly expansionist (Clark, 1982b) and argued that a massive supply of new houses would mop up outside demand and keep house prices low for local residents. This policy resulted in a 45 per cent increase in the number of houses built between 1951 and 1956 but it did not lower house prices, since a large proportion of houses were bought either as second or retirement homes (estimates vary from 26 per cent for holiday homes to 40 per cent for retirement homes), and increasing concern was expressed at the effect of such a rate of building on the landscape.

Therefore in 1974 the policy was dramatically reversed, and the rate of housing development severely restricted so as to preserve the landscape and character of the Park. Only 70 houses a year were to be built, the minimum needed for local employment, reduced household size and demolition. A second thrust of the new policy was the use of Section 52 Agreements under the 1971 Town and Country Planning Act, limiting the use of any houses actually built to local people. The first result of the policy was that although planning applications remained at the same level, the number of applications granted permission fell from 721 in 1972/73 to 143 in 1980/81.

However, it has been argued that this policy will have the direct opposite effect to that intended, that it will again accentuate the problems of the poor and young in the Park (Shucksmith, 1981) and that it will merely continue to divert the attention of house buyers to the existing stock, and thus further inflate the prices of older houses (Shucksmith, 1980a), and so Shucksmith agrees with Clark (1982c) that a better solution may be the development of more subsidized council or housing association housing. A further problem with the policy is that Section 52 Agreements are meant to be limited to land use planning conditions, and it was for a time unclear if they could be extended to social conditions. The matter was clarified in 1983 when the Secretary of State for the Environment ruled that the use of Section 52 Agreements in the Lake District was illegal (or in legal jargon *ultra vires*).

Ironically however, another piece of legislation, the 1980 Housing Act, has introduced a power to control the resale of any council houses sold to sitting tenants, to outsiders, in National Parks, Areas of Outstanding Natural Beauty, and 'designated rural areas' (for a map of the designated rural areas see Phillips and Williams (1983b, 90)). The need for this power had already been illustrated by Gillon (1981) who in a study of council house sales since 1976 in south Oxfordshire, found that it was the most attractive council houses that were sold, and that they were sold to the better-off tenants, leaving a declining council

Table 4.8: Housebuilding in the countryside

New houses built per 1,000 population per annum	Rural districts		England	
	1949–74	1980–82	1949–74	1980–82
Private developer	4.67	3.64	2.67	2.24
Local Authority	1.55	0.77	2.18	1.05
Other public	0.27	0.09	0.38	0.45
Total	6.50	4.48	5.23	3.73

Source: Clark (1984).

housing stock (exacerbated as shown in Table 4.8 by a very low rate of new council house building in the 1980s) to be lived in by the poorest sections of rural society.

The run down of the council housing sector also exacerbates the problems of not only finding a council house, but of transferring from one place to another, and although Phillips and Williams (1982a), in their survey of council housing in south Devon, found that 87 per cent of people on the transfer list and 67 per cent of people on the waiting list had been given their choice of parish, there had been a tendency to concentrate people into fewer locations and into the larger villages. Furthermore, the few council houses that had been built had been overwhelmingly located in the small towns and larger villages of the area.

One major reason for this concentration of building is the economy of scale, since an estate of council houses in a large village is usually cheaper to build than one or two houses in a smaller village. This therefore means that if the people who are least likely to be treated well by present policies are to do better (Phillips and Williams, 1982b, 149), namely 'persons wishing to be housed or rehoused in the smaller settlements' then alternative policies will have to be pursued.

Such alternatives are provided by housing associations, self-build schemes and co-operatives. However, the major thrust is provided by voluntary housing associations funded by the Housing Corporation, but Richmond (1983) has found, in Devon at least, that they provide a highly uneven distribution of houses, and are therefore unlikely to make a comprehensive contribution to rural housing problems.

One final problem area is the tied farm cottage under which a farm worker is given a free or low-rent house, but only as long as he is employed. Although it is difficult to estimate the number of tied cottages, it is generally agreed that between 55 and 60 per cent of farm workers live in the estimated 100,000 tied cottages (Gasson, 1975). It has also generally been agreed by most of the workers in the field that the system should be modified or even discontinued. For example, Irving and Hilgendorf (1975) in a random sample between 1973-74 found a threat of 1,500 possible cases of eviction and 300 probable such cases, and Jones (1975) found that there were few alternative accommodation options for those evicted from tied cottages, and that the system could be a cause of homelessness.

These findings led to calls for abolition, notably by the National Union of Agricultural Workers, and although the 1976 Rent Agriculture Act gave improved security of tenure to workers and their relatives, and also obliged local housing authorities to rehouse a worker where a tied cottage was needed for another worker in the interests of good agriculture, many problems still remain, and Phillips and Williams (1984, 126) conclude that the legislation may 'have done relatively little to help tied cottagers or to improve low-cost agricultural housing'.

This conclusion could easily be extended to other areas of rural housing policy, and given the anti-bureaucratic climate of recent years, Thurgood (1978) has argued that solutions will have to be found within the scope of present powers. The sort of solutions that Thurgood then advances take us back to the arguments for less restrictive key settlement policies, and a greater emphasis on the socioeconomic issues advanced by Cloke (1983) raised earlier in the chapter. Taken to its logical conclusion this could lead to the possibility of housing and planning departments being merged (Phillips and Williams, 1983b) and to the adoption of more flexible policies based on local need.

In conclusion, this chapter has shown that there are definite links between the theory of rural settlement, rural settlement policies and rural housing, but that there has at the same time been a lack of integration between them. Two major themes have also been the concentration of population into the larger villages, and the increasing social segregation of the countryside into owner-occupiers and tenants. The next chapter not only develops these two themes, but also considers how rural populations and communities have changed, and how they are employed in the 1980s.

5

Rural population and employment

Although the distribution of rural population is no longer synonymous with agricultural employment, most people live in the same locality as their place of employment, and so this chapter considers population and employment as contiguous rather than necessarily related phenomena. Within these phenomena four main areas of study have been of interest to rural geographers; first, and most enduring of all, the study of *rural population change*; second, and more recent in its concern, the study of *rural communities*; third, and of more general relevance to other chapters in this book, the *classification and definition of rural areas* by both their population and employment characteristics; and fourth, the study of *employment in rural areas*.

The study of both rural population and employment, however, suffer from the difficulty of finding enough up-to-date data. For example, the main data sources on population (and often employment) are national population censuses, but these are traditionally held on a decennial basis, usually for cost reasons (Rhind, 1983). For instance, the 1980 US Census cost $990 million (Cruickshank, 1981), and mid-term censuses in the UK for both 1976 and 1986 have been cancelled on cost grounds. Nonetheless, many census agencies publish annual, quarterly, and even monthly estimates of population change (OPCS, 1984), but their usefulness is much reduced by the coarse spatial framework in which they are published. For example, the monthly estimates are available at county level only.

Another drawback is that many census estimates concentrate on natural change (i.e. the difference between births and deaths) and tend to ignore the most effective component of population change in the western world, migration. Accordingly, a number of other methods have been employed, for example, lists of those registered to vote (Edwards, 1973) and records of doctor's patients (Rees and Rees, 1977). However, these too have drawbacks, notably the lack of children in the electoral register, and the difficulty of tracing migrants who have migrated to either new doctors (Devis, 1984) or to new electoral districts, but nonetheless they can still pick out new trends, like the repopulation of rural areas, long before the decennial census can (Kennett and Spence, 1979).

Notwithstanding the difficulty of data collection, a coherent body of fact and theory has emerged to produce a number of texts on Population Geography but as (Gilg, 1983c, 75) notes: 'few if any of these studies . . . look at rural populations specifically'. Furthermore, two of the most recent texts (Jones, 1981; Woods, 1979) concentrate on the processes of mortality and fertility and devote only one chapter each, out of nine in both cases, to internal migration. Yet it is migration to and from rural areas that has been the main concern of rural geographers in the last century, and so attention is now turned to the first major theme of this chapter, rural population change.

Rural population change

Although rural populations have traditionally been more fertile than their urban counterparts, by far and away the most significant factor in rural population change over the last century has been migration. Accordingly, a number of migration theories and general statements have been formulated over the years, as shown in Table 5.1. Out of these, push factors 2 and 5 and pull factors 1, 2 and 3 have been most important for depopulation in the western world, and pull factors 4 and 6 have been most important for people moving from urban to rural areas. In a theoretical treatment Woods (1982) divides migration theory into the attitudes and behaviour of both individuals and groups, and then proposes a three-fold structure for integrating theories of migration: First, the theories must have the ability to understand the ways in which individuals form images of their worlds, how they assess those images and then decide whether to migrate or not. In this case aggregate theory tells us that economic criteria are of prime importance in this image building. Second, it must be possible to transfer these theories of group behaviour to generate predictive models, and third to apply this aggregate theory to the behaviour of individuals.

Finally, Mabogunje (1970, 16) has proposed a systems approach which rejects the push-pull approach of Table 5.1 and sees rural–urban migration no longer

> as a linear, uni-directional, push-and-pull, cause–effect movement but as a circular, interdependent, progressively complex, and self-modifying system in which the effect of changes in one part can be traced through the whole of the system.

Empirically however, the last century has been dominated by a major movement of people away from the countryside, although in the past 10 or 20 years there has been a significant repopulation of rural areas. It is therefore prudent to examine both periods in turn before attempting to produce any general statements about rural population change.

Rural depopulation

As long ago as the late nineteenth century Ravenstein (1885) had formulated seven general laws of migration which placed a good deal of emphasis on rural

---

Table 5.1: Summary of factors likely to cause migration

Push factors
1   Decline in a national resource or the prices paid for it e.g. agricultural surpluses;
2   loss of employment from mechanization, e.g. Mechanization in agriculture;
3   discrimination on political, religious or ethnic origins;
4   alienation from community;
5   lack of marriage or employment opportunities in the local community; and
6   retreat because of catastrophe.

Pull factors
1   Superior opportunities for employment elsewhere;
2   opportunity to earn larger income;
3   opportunity to obtain higher education or specialized training;
4   better living conditions;
5   dependency, moving with marriage partner or to become married; and
6   wish to enter a new sort of community.

Source: Lewis (1982).

depopulation. For example: Law 2 argued that urban areas will absorb the population of first their immediate rural areas and then even the remote areas; Law 6 stated that urban dwellers are less migratory than rural inhabitants; and Law 5 argued that migrants proceeding long distances generally migrate to the great centres of commerce and industry. However, Ravenstein in Laws 1 and 4 also pointed out that most migrants only move a short distance (Law 1) and that each main current of migration produces a compensating counter-current (Law 4).

Until the recent repopulation of the countryside this last law, Law 4, didn't seem to apply and Laws 2 and 5 seemed to be operating inexorably. For example, during the first half of this century, Saville (1957,7) noted that the percentage of the population in urban districts has risen from 72.0 per cent in 1891 to 80.7 per cent in 1951 and that:

> The basic cause is everywhere the same. Rural depopulation has occurred in the past century and a half, and will continue in the future, because of declining employment opportunities in the countryside. Economic activities have steadily moved from the villages and the rural communities into the towns and the urban areas; and as employment opportunities have diminished in the rural areas, the village population have moved into the towns.

Nonetheless, Saville also pointed out that the rate of decline had begun to slow down, and this was also confirmed by other studies of the time. For example, Vince (1952) concluded that the primary population had fallen both in numbers and relative proportion during the period 1921–31, but that in some areas a new 'adventitious' population was beginning to replace them. Willatts and Newson (1953) confirmed that depopulation trends had continued till 1951, but that some rural areas were by then beginning to show signs of recovery. Nonetheless, the recovery was largely confined to the more prosperous southeastern lowlands and also, within these and other areas, to the larger villages and small towns.

For example, Bracey (1958) in a study of 375 Somerset parishes found that, in general, the more remote and less well serviced parishes were those with the worst and most persistent depopulation, and findings like this only encouraged the further development of theories of settlement concentration in the 1960s, although these can also be traced back as far as 1918, when Peake (1918) advocated equally-spaced villages with populations of between 1,000 and 1,500 people.

Indeed in the 1960s, other reasons for depopulation, notably settlement size, began to be examined. For example, Turnock (1968) in a study of northeast Scotland found that depopulation in the area was not only a function of a retreat from hill farming but also a process of concentration into fewer but larger settlements, and Johnston (1965) in a study of northern England found that village size was important in not only retaining the existing population but also attracting newcomers from nearby towns. These findings were also confirmed in the USA, where Butler and Fuguitt (1970) and Hassinger (1957) from work in Wisconsin and Minnesota, found that rural settlements near larger albeit not very much larger towns, tended to grow faster than average, although the findings were not universal through either time or space.

Other reasons for depopulation were also examined during this period. For example, Hannan (1969) found that the number of young people in an area tended to increase outmigration, as the more able young left to follow the kind of careers not available in rural areas, though Grafton (1982) has argued that young people do not outmigrate from remote rural areas at a faster rate than their counterparts in less remote rural areas, and that any decline in such areas is due to lower levels of immigration, rather than higher rates of outmigration. Other work has examined kinship, and Johnston (1971) has found that this has tended to reduce migration. In another study of migrants in central Wales, Jones (1965) even went so far as to test all of Ravenstein's laws, but he could only definitely confirm Laws 1 and 5, which argue that most migrants only travel a short distance, but that those who do travel further migrate to the great centres of commerce and industry.

In summary, the detailed empirical work outlined above has enabled another general law to be formulated, the principle of circular and cumulative causation, in which the key factors are: an initial reduction in farm employment; which is followed by a deterioration in the age structure; which is in turn followed by a reduced rate of natural increase which then leads to a reduced population which in turn leads to reduced demands for rural services and thus a reduced demand for rural employment which then triggers another circular decline (Hodge and Whitby, 1981). However, the rural repopulation trends, already noted earlier in the chapter, began to gather pace in the 1960s, and by the 1970s and 1980s, a new phenomenon emerged which seriously undermined the process of decline namely the phenomenon of counterurbanization.

Counterurbanization and rural repopulation
This new phenomenon wasn't quickly recognized in the literature and as recently as 1980, White and Woods's edited text (1980), *The Geographical Impact of Migration*, included three lengthy index entries on rural depopulation, rural–rural migration and rural–urban migration, but nothing on rural repopulation or counterurbanization, although there were four index entries for return migration. In another example, Lewis (1982, 177) limits the repopulation to those 'rural areas accessible to metropolitan centres and small cities'.

One reason for this slow recognition may have been doubt about the term itself and confusion concerning what is actually meant (Dean *et al.*, 1984a). What wasn't in doubt, however, was the process itself, and even for the 1961–71 period, Champion (1976) was able to conclude that a marked acceleration had taken place in the rate of outward movement, from the conurbations and large towns, to surrounding areas, and that remote rural areas like the southwest and Wales had either reversed a century-long decline in population, or had reduced it significantly. These trends continued and accelerated in the 1970s, and are discussed more fully in the summary at the end of this section, but in advance of the more definitive work allowed by the 1981 Census, a number of local studies began to identify a group of common themes.

For example, Jones (1976) in a study of central Wales found that employment, the desire to live in a better area and a former link with the area, were important reasons for explaining the success of a growth pole town, Newtown. In a more recent study in West Cornwall, Dean *et al.* (1984b) found that the immigrants tended to be older, of a higher occupational status, from the more prosperous regions and to have former links with the area. Indeed, 22.4 per cent of the migrants could be defined as return migrants (Shaw, 1984) and although this isn't as high as the 30 to 40 per cent rates recorded in north Norway

(Nicholson, 1975) it is much higher than the 5.5 per cent found in northwest Ireland (Foeken, 1980), and very different from the situation in northern Scotland where Jones *et al*. (1984) have found that a significant amount of the immigration was from English-born people rather than from return migrants.

Many return migrants are elderly, and indeed retirement migration is a major factor behind counterurbanization in many rural areas, notably coastal and scenically attractive areas, and in all likelihood this trend will accelerate, as both the percentage of elderly in the population increases, and the age of retirement continues to fall (Law and Warnes, 1976). By itself this might not be a bad thing, but elderly inmigration has tended to be accompanied by youthful outmigration, leading to a deteriorating age–sex structure. Furthermore, counterurbanization may not necessarily reduce youthful outmigration, since although surveys in both prosperous and marginal regions (Drudy and Drudy, 1979) have shown that 49 per cent in the prosperous areas wanted to leave, compared to 68 per cent in marginal areas, in both areas, around 30 per cent of school leavers aspired to the sort of professional job not traditionally found in rural areas.

Another reason for the relatively slow acceptance of the term counterurbanization in Britain may be the fact that Britain may have been one of the last developed countries to experience the phenomenon. For example, in the USA, books on the topic appeared as long ago as 1980, when Brown and Wardwell (1980) made the following observations about the phenomenon:

1   For the first time this century, population and economic growth in non-metropolitan America is exceeding population growth in metropolitan America.

2   Growth is occurring in remote and completely rural counties, as well as in counties that are partly urban or dominated by nearby metropolitan centres, but the south and west are increasing the fastest.

3   In most regions, the new growth is entirely due to changes in net migration, as movement out of the conurbations has exceeded the inflow to these areas, and low or convergent rates of natural increase have provided a passive background for these trends.

4   The migration reversal is as pervasive across the socio-demographic characteristics of movers as it is across regions.

In another study, Johansen and Fuguitt (1984) from a sample of 572 of the 11,334 villages in the USA (defined as a population of less than 2,500 in 1960) found that while only half of the villages had grown in the 1950s, two-thirds had grown in the 1970s, and that this growth was no longer restricted to the larger villages.

In spite of these broad generalizations the exact reasons for the turnaround aren't very clear, although Brown and Wardwell (1980) have suggested that three interrelated factors are economic decentralization, a preference for rural living, and the modernization of rural life. But Campbell and Johnson (1976) have also concluded that a large amount of research needs to be done in this area. Unfortunately, most of the research conducted so far hasn't produced more than a few tentative generalizations. For example, De Jong and Humphrey (1976) found that the migrants tended to be younger, of a higher socioeconomic status and lived in smaller households, while Wardwell (1977) concluded that one major motive was a desire to live in smaller settlements and communities; but rather surprisingly, Fliegel *et al*. (1981) found that this type of migrant wasn't opposed to further growth in the new neighbourhood.

What is clear, however, is the widespread nature of the trend, not only in the USA which the 1980 Census confirmed, as shown in Figure 5.1 (Cruickshank, 1981), but also in Europe where Fielding (1982) concluded that in nearly all the countries of western Europe counterurbanization has replaced urbanization. It also seems clear that the trend will soon emerge elsewhere, for as Vining and Pallone (1982) in their study of 22 countries have argued, the phenomenon is due to the fact that the scale economies of the core regions of any country either have been or are being reduced, as the peripheral areas manage to offer sites that are as competitive as those of the core regions.

Nonetheless, in spite of the widespread nature of the phenomenon, there is no agreed explanation of the causes, and these are probably as complex as the process itself, which is by no means as simple as it first seems. For example, Fielding (1982) takes the best elements of the three main models: (a) the desire to live in a more rural setting; (b) the availability of new jobs in the sunrise industries of the rural areas; and (c) regional planning by governments to argue that the two main factors are first, the development of a post-industrial society where the mobile middle classes are relocating the new industries, and second, the rise of service employment, particularly in the more attractive rural areas of each country.

However, Dean *et al.* (1984a) have criticized both Vining and Pallone's and Fielding's approaches as being too 'logically positivist' in their approach. Moreover, they argue that the term counterurbanization poses four significant problems. First, there are problems over 'ambiguity', and whether to use Fielding's definition of counterurbanization as the opposite of urbanization, or to extend the term to Robert and Randolph's (1983) prerequisites including both decentralization and deconcentration. Second, the term 'presupposes' the very phenomenon it seeks to identify. Third, there is already a process of 'reification' although the process may be more apparent than real. Fourth, and finally, although the process has been seen as an international trend, most studies have been too 'parochial' and 'logically positivist', and even in Fielding's work (Dean *et al.*, 1984a, 10) which: 'breaks out of the consensus mould and moves towards neo-marxist interpretations, a certain narrowness of outlook remains'. Therefore Dean and his colleagues argue for a less 'positivist' and a more broadly based research effort, including marxist and structural approaches.

However, this type of work immediately meets the problem of data collection outlined at the beginning of this chapter, and for the foreseeable future most research will have to rely on official census data with all their limitations. Nonetheless these data have allowed both the descriptive work outlined in the next section and the explanatory work outlined in the section after that (on the 'Classification and definition of rural areas') to be produced.

Rural population change: A summary

The first point to be made about recent rural population change is that the broad spatial processes of the 1970s and early 1980s can be traced back into not only the 1960s, but even the 1950s (Compton, 1983). Compton has also concluded that decentralization has continued despite the energy crisis, the south is still gaining population at the expense of the north; East Anglia, the South West and the East Midlands continue to be the fastest growing regions; and that Wales is

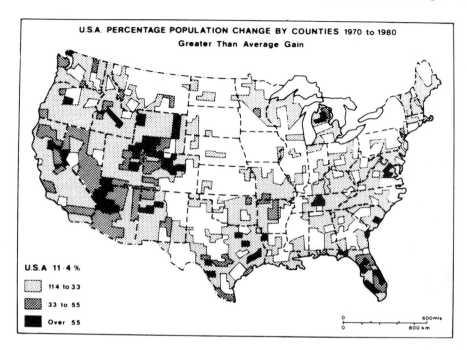

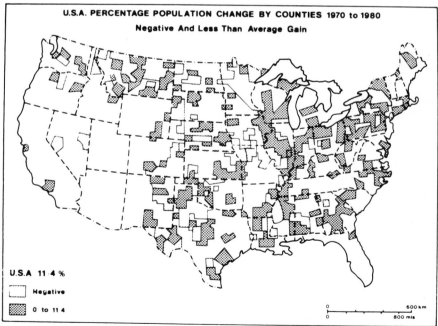

Fig 5.1 Population change by counties in USA 1970–80. Source: Cruickshank (1981)

about to join them. However, the 1970s and early 1980s have seen one major change in population, not foreseen in the 1960s, a massive drop in the birth rate and a virtual cessation of overall population growth, so that the increase of 5.4 per cent recorded in 1961–71 fell to only 0.5 per cent in 1971–81.

This means that migration now has a major impact on population change, even though Ravenstein's first law is still seen to hold, since Brant (1984) has shown that of the 9 per cent of people who moved between 1980 and 1981, a massive 70 per cent moved less than six miles. The most mobile age group in this period were those aged 16 to 34, and the most popular destinations were the so-called 'sun-belt' counties of Wiltshire, Oxfordshire, Surrey, Suffolk and Cambridgeshire.

Within these rural areas there seems to be a definite preference in favour of not only smaller towns (i.e. those below 50,000 people) but also villages, for both the periods 1961–71 and 1971–81, and for both the USA and England and Wales as shown in Table 5.2. However, the process is by no means uniform and Table 5.3 shows that a small number of towns in the rural districts actually lost population betwen 1971 and 1981, although the dominant trend was increases of over 10 per cent in both of the rural categories at the foot of the table.

This is demonstrated spatially by Figure 5.2, which shows population change by aggregated districts, and although districts aren't the most suitable way of portraying population change, three areas of rural growth stand out. First, the so-called 'sun-belt' from Norfolk through Oxford to Bournemouth, with many districts showing a growth of 15 per cent and over, second, a broad band of growth between 5 and 15 per cent across all of lowland England and the Welsh Marches, and third, a significant area of growth with rates of 30 to 60 per cent (Carruthers *et al.*, 1984) in the Scottish Highlands, but in this case it must be remembered that the absolute totals are fairly small. The only areas to show only modest growth are west Wales, northern England and southern Scotland, where hill-farming environments have failed to attract either sunrise employment or the early retired.

Table 5.2: Population change by size and type of settlement

England and Wales 1961–71 Settlement category	Population change	Rate %
Conurbation	− 813,910	− 4.9
Urban areas above 100,000	+ 113,657	+ 1.7
Urban areas between 50,000 and 100,000	+ 373,509	+ 7.5
Urban areas under 50,000	+ 1,201,478	+ 13.7
Rural districts	+ 1,614,376	+ 18.0
England and Wales	+ 2,489,110	+ 5.4

USA 1960–1970 and 1970–1980 Settlement category	% change 1960–70	% change 1970–80
Metro	+ 17.0	+ 9.8
Non-metro	+ 4.4	+ 15.8
Non-metro		
Adjacent counties	+ 7.3	+ 17.4
Non-adjacent counties	+ 1.4	+ 14.0
With city of above 10,000	+ 7.0	+ 14.5
With no city of 10,000	− 2.7	+ 13.6
Total	+ 13.4	+ 11.4

Sources: Champion (1973); Beale, (1982).

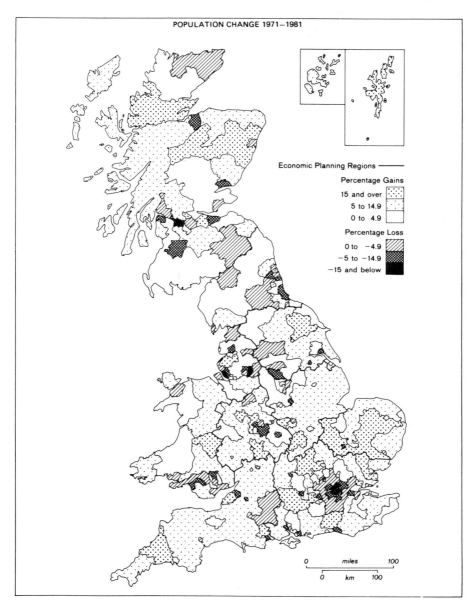

Fig 5.2 Percentage population change for Great Britain 1971–81 by district council areas.
Source: Randolph and Robert (1981)

Table 5.3: Population change by type of district

Category of district	1981 millions	No. of districts	% change 1961–71	Change 1971–81 '000	% change 1971–81	Decrease Over 5%	Decrease 0 to 5%	Increase 0 to 5%	Increase 5 to 10%	Increase Over 10%
England and Wales	49.0	403	5.7	262	0.5	55	71	92	88	97
Inner London	2.5	14	– 13.2	– 535	– 17.7	13	—	—	—	1
Outer London	4.2	19	– 1.8	– 221	– 5.0	9	10	—	—	—
Major cities	3.5	6	– 8.4	– 386	– 10.0	5	1	—	—	—
Other cities	7.7	30	5.5	– 160	– 2.0	7	11	12	—	—
Other cities over 175,000	2.8	11	– 1.4	– 149	– 5.1	4	6	1	—	—
Smaller cities	1.7	16	2.2	– 55	– 3.2	6	5	5	—	—
Industrial districts	6.7	73	7.9	200	3.3	6	18	24	16	9
New Towns	2.2	21	21.8	283	15.1	•1	1	4	3	12
Resort and sea-side retirement	3.3	36	12.2	156	4.9	2	3	11	12	8
Outer urban, mixed urban-rural, and more accessible rural districts	9.4	99	22.0	661	7.8	2	14	21	28	34
Remote largely rural districts	5.0	78	9.7	468	10.3	—	2	14	29	33

Source: OPCS (1981a).

In spite of the fact that Champion (1981, 20) has argued that 'some of the most spectacular changes have occurred in those rural areas which are relatively remote from traditional metropolitan influences' it must also be remembered that the extra numbers in the remote areas are relatively modest, compared with those in the outer margins of the main centres of population, as shown in Figure 5.3, and that most of the population growth has been strongly associated with smaller towns and accessible settlements in the countryside both in England (OPCS, 1981b) and Ireland (Duffy, 1983). If this is the case, the changes of the past 30 years may be the first signs of a return to the more traditional population distribution of pre-industrial Britain, but it must also be borne in mind that it takes a very long time to shift major population patterns, and that the present trends may only be a veneer on an underlying and more permanent structure.

In conclusion, Robert and Randolph (1983) from a major study of the 1961, 1971 and 1981 Census data conclude that the counterurbanization explanation may not be as appropriate as first thought, and that the crucial flaw in the model is that (1983, 97) it is 'a model of pattern rather than process'. However as Pacione (1984, 145) observes: 'few models have been constructed to explain the population turnaround', but out of these the model produced by Lewis and Maund (1976) and shown in Figure 5.4 may fit the present situation as described by Moseley (1983) in Table 5.4 the most closely.

Whichever explanation is revealed by the major research effort that is still needed for the rest of the 1980s, the population changes discussed above have also had a major impact on rural communities, and so it is to this topic that attention is now turned.

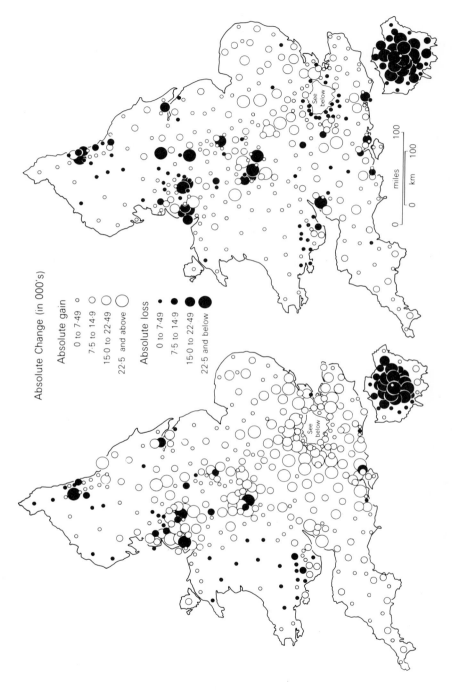

Fig 5.3 Absolute population change for England and Wales 1961–71 (left) and (1971–81) right. Source: Robert and Randolph (1983)

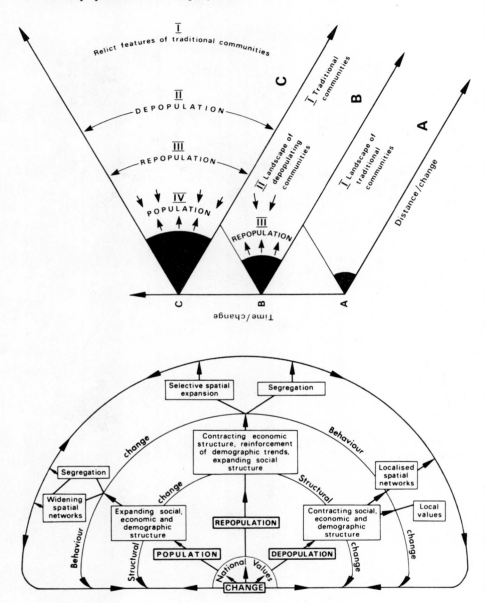

Fig 5.4   Two related models for urbanization and population change in the countryside. Source: Lewis and Maund (1976) (a) Upper diagram (b) Lower diagram

Table 5.4: Summary of rural population change

Depopulation has not been the norm in post-war rural Britain.
Broad figures from the pre-1974 rural districts suggest a rural population of 10.4 million in 1951, 12.8 million in 1971 and an estimated 14 million in 1981, with percentages of 20 per cent in 1951, 23 per cent in 1971 and 25 per cent in 1981. Within these broad themes, four distinctive trends can be outlined;

A   *Depopulation of the remotest areas*, especially of young people, generated by a marked decline in employment opportunities and to, a lesser extent, a dissatisfaction with the local social and cultural amenities;

B   *Urban decentralization*, linked in part to the dispersal of employment from the inner areas of the large cities, and in part to the search for better, cheaper and better-situated housing;

C   *Retirement migration* (and the 'seasonal extension of suburbia' into second homes) in which late-middle-aged and elderly people seek cheaper housing in attractive rural areas; and

D   *Local restructuring*, in which population change within rural areas is positively related to the size of villages and small towns: the largest settlements grow at the expense of their smallest neighbours.

Source: Moseley (1983).

Rural communities

Because of the population turnaround in many rural areas, the definition of a rural community in a widespread and confused literature has become an elusive concept (Lewis, 1979) and although most parts of the UK have reached Time C in Figure 5.4a, some areas are characterized by a mixture of repopulation and depopulation in Zone III, and depopulation and repopulation in Zone II, so that the hallmarks of both the population and depopulation quadrants of Figure 5.4b can be found together in a very small area. Another problem according to Lewis (1983, 149) is 'the lack of an adequate conceptual framework' and as a result (1983, 149): 'most rural studies have tended to be empirical and holistic, thus lacking comparability and depth'.

This was certainly true of the majority of studies before and even till the 1970s, which ususally adopted a social anthropological approach (Arensberg and Kimball, 1940) to static or declining farm communities (Davies, 1970) and even by 1971 it was possible for Franklin (1971, 12) to argue that 'the peasantry, though diminishing in number and importance, still forms the largest single category within the rural community throughout Europe'. However, the effects of urban growth on rural communities had been recognized as long ago as the late 1930s, when Ashby (1939) argued that the unbalanced age structure and uneconomic service provision caused by rural depopulation had been offset by the spread of urban population into the rural areas in the inter-war years.

By the early 1970s, the spread of urban influences and easier communications and telecommunications meant that a single culture began to emerge (Mogey, 1976) with regional and urban–rural differences being much reduced, and this invalidated the century-old division of society by Tonnies (1887) into 'Gemeinschaft' (rural) and 'Gesellschaft' (urban). Instead, the idea of a continuum emerged (Pahl, 1966b) based on the degree of urbanization experienced in an area and in the early 1970s Rogers and Burdge (1972) produced the continuum which is shown below and which depends upon population size, population density and the degree to which the community members observe rural or urban norms.

Rural–Urban Continuum

Rural neighbour- hood	—	Agricul- tural village	—	Small town	—	Rural–urban fringe community	—	Suburban community	—	Small city	—	Metro- politan city

Rogers and Burdge (1972, 264) have also provided a definition of a community which provides a useful insight into why so many people have moved into the countryside in recent years, since it reads:

> A community is a group . . . it is composed of people in communication and together geographically with common interests or ties. A community is simply one kind of group in which membership is based on locality.

Since the locality in the case of a rural community is a village, with obvious territorial limits in contrast to the more amorphous boundaries of suburban neighbourhoods, many people have migrated to the countryside in search of a more friendly and obviously defined community. Not surprisingly, the social impact of these migrants on rural communities was a major topic of study in the 1960s and 1970s, but in the 1980s, a new theme, related more to the political impact of the migrants on the power structure of rural communities has emerged, as rural sociologists and geographers have sought a more critical and a less apolitical and anodyne approach to rural society. Accordingly, the rest of this section concentrates on first, studies of social change in rural communities, and then, the changing political structure of these communities, before considering one model for the future, California.

Social change in rural communities

Studies of social change in rural communities have been a major theme in rural community work since the early 1950s, when the first typologies (some of which have already been discussed in Chapter 4, and notably in Tables 4.6 and 4.7 were developed. For example, Mitchell (1950; 1951) proposed two typologies as shown below:

Mitchell's 1950 typology	Mitchell's 1951 typology
Integrated open	Suburban villages
Integrated closed	Squirearchy
Disintegrating open	Structureless
Disintegrating closed	Rapid depopulation, decaying structure

Thorns (1968), from empirical work in Nottinghamshire, defined three types of village: first, villages with an established patten of stratification based on the traditional rural economy or squirearchy; second, villages in a state of transition; and third, re-established villages where professional newcomers have taken over the dominant roles. The most influential typology however was probably that produced by Pahl (1966a). already outlined in Table 4.6, in which the most important groups were the 'spiralists' and 'council-house tenants'.

This typology further developed the fourfold classification of the characteristics of changing communities in the rural–urban fringe, outlined in Pahl's (1965) classic text *Urbs in Rure* namely:

1 *Segregation* of rural areas into large blocks of one class or price of housing;
2 *Selective inmigration* as rural-urban fringe parishes attract mobile, middle-class commuters who live and work in distinct and separate social and economic worlds from the established populations; and although
3 *commuting* is not confined to the middle classes the working classes are more limited by public transport; and
4 *the collapse of geographical and social hierarchies* as the community becomes more outward looking and the squirearchy is replaced by a class-based structure polarized around housing segregation.

Much subsequent work has tended to confirm both Thorns's theoretically based and Pahl's empirically based conclusions. For example, Radford (1970) in a study of two contrasting Worcestershire villages found one village with an established pattern of stratification, but in the other village with a strong newcomer element, she found an increasingly mobile and polarized society. In another study, this time of one village Ringmer in Sussex, which had doubled in size from around 2,000 inhabitants in 1961 to 4,000 in 1971, Ambrose (1974) found that the main reason for migration to the village had been for a job, or to be within commuting range of a job, thus confirming that the main reason for migration, as already outlined earlier in this chapter, is economic, but also that the most dominant newcomers are Pahl's 'spiralists'. However, Ambrose also found that these 'spiralists' did take part in community life, and didn't see the village as a transient bus stop as Pahl had forecast.

Nonetheless other work has confirmed Pahl's view that inmigration would lead to social segregation. For example, Connell (1978, 131) in his study of central Surrey concluded that the growing division between the owner-occupiers (middle class) and council tenants (working class) 'characterizes many of the social relationships' within the area. Perhaps the most compelling confirmation has however come from Pacione (1980) and his study of the 'metropolitan village' of Milton of Campsie to the northeast of Glasgow. Like Ringmer, the population of the village has rapidly expanded in recent years from just under 2,000 in 1971 to over 3,000 in 1981. Furthermore, most of this growth has taken place in the large but separate estates of owner-occupied and council houses outlined by most typologies. It is therefore in many ways an ideal empirical test bed for assessing the validity of Pahl's classification in the 1980s. Some of the results are shown in Table 5.5, and Table 5.5a confirms the much more mobile nature of the private households. Table 5.4b confirms the social-class divide between private and council households, and Table 5.4c confirms that the village is socially divided but that the middle class are far more open in their social activities, both socially and spatially. In summary, Pacione's work appears to confirm Pahl's (1966a, 1149) conclusion that:

> To the working class, life in the village means family life and possibly life at work, to the middle class spiralists, life in the village means life in the voluntary associations.

Nonetheless, it would be unwise to conclude that all areas or villages are falling into Pahl's typology. For example, many villages haven't experienced the inmigration of young professional commuters, and in many areas, like Dorset, young adults are moving out and retired people are moving in (Moss,

Table 5.5: Socioeconomic characteristics of life in a metropolitan village

a. *Migration distances*	Private households		Council households	
Place of origin	No.	%	No.	%
Village born	0	0	17	37.0
Within 5 miles	13	17.8	22	47.8
Within 20 miles	26	49.3	3	6.5
Rest of Scotland	9	12.3	0	0
Great Britain	12	16.4	3	6.5
Other	3	4.2	1	2.2

b. *Socio-economic structure*			
Social class	Strathclyde	Milton of Campsie	
1971	region	Private	Council
I	10.3	25.7	0
II	36.1	39.2	16.7
III	28.2	32.4	63.9
IV	14.0	2.7	8.3
V	11.4	0	11.1

c. *Social activity patterns*	Private households		Council households	
Area	No.	%	No.	%
Within street	35	32.4	18	24.7
Within class	31	28.7	43	58.9
Across class groups	5	4.6	5	6.8
Outside village	37	34.3	7	9.6

Source: Pacione (1980).

1980) leading to a quite different set of social problems. In the most remote areas, like central Wales, it is still possible to find three traditional types of rural community namely 'scattered settlements', 'rural villages' and 'market towns' (Williams, 1976) and in other rural areas, like North Herefordshire, Lewis and Maund (1979) have questioned the true significance of social class, life style and length of residence as determinants of intra-community behaviour. However, Lewis and Maund (1979) did find that social segregation was occurring in the areas with the largest influx of newcomers and the largest populations.

One other facet of rural community change is provided by the fourth element of Pahl's classification: the collapse of geographic and social hierarchies. These have been best documented by Newby (1979) and his colleagues working in East Anglia, and though East Anglia is by no means typical of rural Britain, there is little doubt that their findings are probably fairly universal. First, they confirm that rural life has become socially polarized by the influx of newcomers, second, that the traditonal rural economy has been eclipsed, and third, that the social power of the squirearchy has been much reduced (Newby *et al.*, 1978). However, they do not agree that the political power of the farming community has also been eclipsed, and they found that most local councils were still dominated by farmers (Newby *et al.*, 1978). Furthermore they concluded that farmers had used this political power to prevent new employment in order to keep farm wages lower, and to keep rural council house building rates at a minimum so as to keep farm workers tied to their cottages. Nonetheless, they also found that the professional newcomers to the area were often upset by the environmental consequences of modern farming (see Chapters 8 and 9) and since they lacked a rural power base they formed environmental pressure groups to protest against the farmers, thus creating a new form of social division in the rural community.

Although Newby and his colleagues have demonstrated the dogged hold of traditional rural society on the levers of power, the changes foreseen by Pahl and Thorns, and documented by Ambrose, Connell, Radford and Pacione, may be only the foretaste of a much greater change in rural society, if the development of rural communities in California is a guide to the future as it has been so often in the past.

In California the advance towards a new 'post-industrial' society has been more rapid than elsewhere and according to Bradshaw and Blakeley (1979,6) the four distinguishing characteristics of this 'prototypical advanced industrial society' are: (a) high technology (for example, the explosive growth of micro-computers in the hitherto rural 'Silicon Valley'); (b) up to 70 per cent employed in the service (tertiary) sector and only 5 and 25 per cent respectively in the primary (agriculture) and secondary (manufacturing) sector; (c) the intense rate at which knowledge is generated and transmitted; and (d) increasing interdependence between places for goods and services. To some degree these conditions are already being experienced in Britain; in central Scotland, along the M4 corridor, and around some of the New Towns – and if they continue to develop, then Warner's (1974, 315) forecast will certainly come true:

> The challenge of a post-industrial society for persons interested in knowledge about rural society is not the loss of significant work to be done, but the overwhelming scope and complexity of the task before us.

Spatially, this will be expressed in further counterurbanization and as Bradshaw and Blakeley (1979, 27) point out: 'the rural areas [of California] grew at a rate nearly double that of the state as a whole' between 1970 and 1976. Once again if these changes continue to develop in the UK, the classification and definition of what constitutes a rural area will have to be radically altered, although there have already been some changes, as the next section demonstrates.

The classification and definition of rural areas

It has already been pointed out that the definition of a rural area, or a rural community, is an elusive concept, but many authors would agree with Bealer *et al.* (1965) that most understandings involve the use of either *ecological*, *occupational* or *cultural* dimensions. However, the first two approaches suffer from a number of defects. For example, Buttel and Flinn (1977) in a survey of Wisconsin residents found only weak support for the hypothesis that ruralism was more strongly associated with environmental (*ecological*) concern rather than agrarianism. The second approach, *occupational structure*, has been invalidated by both the population turnaround and the loss of agricultural employment already discussed, but it did once allow a reasonable classification to be made consisting of agricultural-rural, rural and rural-urban (Robertson, 1961). This therefore leaves the *cultural* dimension as the most important approach, but as Bealer *et al.* (1965, 266) argue: 'a single dimension . . . would probably not receive widespread acceptance. A composite definition has more overwhelming appeal'.

Not surprisingly, therefore, most approaches have been based on the cultural dimension with the addition of ecological and occupational factors. The simplest cultural dimension is to use local authority administrative areas but these often lag behind significant changes in population composition, and although Robertson's (1961) classification already outlined, defined 75 per cent of the areas as rural-urban, she was also able to comment that this reflected not only the extent of the outward movement from towns, but also the inadequacy of the census definition of rural population, namely all persons living in administrative Rural Districts. However, in the 1981 Census an attempt was made to physically define urban areas, at the enumeration district level, by using a threshold of 1,000 people and 20 hectares of urban land (Denham, 1984). These low thresholds gave an urban population embracing about 90 per cent of the total, and the rural/urban differences shown in Table 5.6, which reveals that the greatest contrasts are to be found in cars per household and the percentage of people travelling to work on public transport (both related to each other and to the low density of rural areas); the percentage in social classes I and II; the rate of unemployment; and the percentage of those working in manufacturing.

Although these areal units will alter less with time than administrative units, they are nonetheless transitory and so, ideally, smaller and more permanent units like kilometre grid squares should be used, as they have been by the OPCS in a limited fashion, as the basic building block. Nonetheless a good deal of basic information can still be portrayed with the traditional concept of population density based on administrative units. For example, density figures can show (OPCS, 1980, 2) that: 'the greater part of the physical area of the countryside is inhabited at relatively low densities'. Put another way, 25 per cent of the population are spread over 90 per cent of the land area, or that while the average density is 2.4 people per hectare, the median density is only 0.5 people per hectare. Nonetheless, the population turnaround and planning controls are imposing average densities over larger areas of the country, and Craig (1979) reveals that it is the areas of suburban density that have experienced the greatest overall population growth in both population and extent.

Best and others (1974, 206) have further developed this theme by using regression analysis to propose a density–size rule which states: 'As the population size of settlement increases, the land provision declines exponentially'. Empirically they found that densities were tending towards a pivotal density, with metropolitan areas shedding their populations by redevelopment and smaller country towns increasing their densities by infilling. In another study, based totally in rural areas, Best and Rogers (1973) used land use maps from the Second Land Utilization Survey and the development plan maps of local planning authorities to produce a least-squares regression model of the form:

$$\log y = 1.9986 - 0.1522 \log x$$
where y = total land provision (ha/1000 people)
and x = population of settlement

This model was able to show deviations from the norm and thus show areas of unexpectedly high or low density.

In spite of this useful work, density remains a rather static and misleading statistic in that it depends very much on the unit used to compute the density. An

Table 5.6: Selected characteristics of urban and rural areas in England and Wales

Population size in '000s	% of pensionable age	% of houses owner-occupied	Cars per house-hold	% of men 16–64 unemployed	% working in manu-facturing	% in Social class I and II	% travelling to work by public transport
All urban areas	17.7	57.3	0.76	11.8	28.2	22.3	29.7
1,000 and over	17.5	52.8	0.72	12.1	26.5	23.1	—
500–999	18.1	46.4	0.60	16.4	30.5	17.2	50.6
200–499	18.4	61.5	0.73	13.3	29.8	20.0	41.5
100–199	17.3	59.3	0.74	11.3	30.6	20.5	34.1
50–99	17.5	58.1	0.76	11.8	29.6	21.4	30.7
20–49	16.9	60.0	0.81	10.9	29.2	23.2	23.6
10–19	17.6	62.6	0.85	9.3	27.8	24.5	21.3
5–9	17.5	62.7	0.88	8.7	27.0	24.7	17.4
2–4.9	18.1	65.5	0.97	7.8	25.3	28.4	18.5
Under 2	19.7	64.7	0.99	7.9	23.4	28.5	18.6
Rural areas	18.5	62.3	1.22	7.2	19.9	34.7	9.2

Source: Denham (1984).

alternative but related statistic, population potential (Craig, 1972) has therefore been developed to take into account the catchment area of population, but once again different spatial units can produce considerable variations in population potential distributions (Craig, 1974). Nonetheless, the concept has been developed to include accessibility and this use is discussed more fully in the next chapter on Rural Transport and Accessibility.

Density might have remained the most accepted form of classification but for three developments in the 1970s which gave a great impetus to the classification of rural populations into different typologies or groupings:

1 The advent of really powerful computers;
2 The publication of census data on computer tapes or discs, and increasingly by smaller spatial units, and even more usefully, grid squares; and
3 The widespread availability of package programs for multivariate statistical techniques like factor, cluster and principal components analysis.

Nonetheless, some of the models developed were still fairly simple. For example, Craig and Frosztega (1976) simply tested whether density declines with distance from the centroid of a population centre, and if the population of these centres follows some regular pattern (e.g. rank-size or log-normal), and found an increasing fit from 1931 to 1961.

Far more complicated were the models based on multivariate statistics. For example, Cloke (1977b) used 13 census variables and three variables based on distance from an urban centre, in a principal components analysis, to produce the index of rurality for 1971 shown in Figure 5.5. Although one suspects that the extreme rural and intermediate rural areas would be reduced in area if the same exercise were repeated for the 1981 census (unfortunately it hasn't and anyway would be difficult to do so since local authority boundaries have changed since 1971), the analysis did allow Cloke (1978) to compare degrees of rurality between 1961 and 1971, and to produce a three-fold typology of rural areas that were becoming increasingly non-rural, were static or becoming

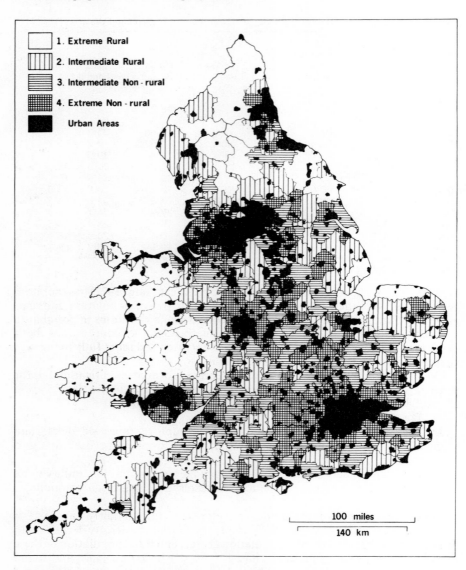

Fig 5.5   Cloke's index of rurality. Source: Cloke (1977b)

increasingly rural, and to produce maps of these three types for each of the four rural areas shown in Figure 5.5. Cloke then proposed a cyclic process for rural change with reverse depopulation and reversed growth as two poles of the cycle. This followed the same lines as Woodruffe (1976), who divided rural districts into six types depending on their history of population change:

1 Accelerated depopulation
2 Reduced depopulation
3 Reversed depopulation
4 Reversed growth
5 Reduced growth
6 Accelerated growth

and Matthieu and Bontron (1973) who divided population change in France between 1962 and 1968 into:

A Maintenance or acceleration of depopulation
B Weak slowing down of depopulation
C Strong slowing down of depopulation
D End of depopulation
E Acceleration of repopulation.

Another example of the multivariate approach is provided by Webber and Craig (1978) who used 40 variables from the 1971 Small Area Statistics Census data to classify the British population into six main family types and a number of sub-clusters, making 30 groups altogether, as shown in a simplified form in Figure 5.6.

The rural and resort areas shown in Figure 5.6 are characterized most of all by a high concentration of self-employed and agricultural workers, and small proportions in the 15–24 age group. Socioeconomically the rural areas are close to the national mean but are over-represented by professional/managerial and semi-skilled workers. This dichotomy betwen rich and poor in rural populations is also brought out by other variables, for example, an above average share of owner-occupied housing is counterbalanced by an above-average share of unfurnished privately rented accommodation (probably tied cottages).

Although these classifications and their spatial pattens tend to confirm the generally accepted view of rural population already discussed here and in Chapter 4, Webber and Craig admit that multivariate classification is a subject on which virtually no two practitioners agree either in the UK or in other countries like France (Chapius, 1973) and therefore argue that the crucial test should not be the methodology but whether the classifications are useful.

Nonetheless, in a critique of the classification, Openshaw *et al.* (1980) point out that statistical classification methods are by their very nature exploratory data-analysis techniques, dependent on the methods used and on a number of arbitrary definitions, and it is therefore inevitable that the results should be seen as a pioneering first attempt, rather than as a polished final product based on the application of well established proven methods. In a reply to Openshaw's strictures Webber (1980), however, reiterated the difficulty of working with such vast data sets and with new complex statistical procedures involving dozens of variables, and claimed that the classification was as far advanced as any could be, bearing in mind the constraints of public policy and the realities of time and cost.

Not surprisingly therefore, other workers have abandoned the census and looked at individuals. For example, Es and Brown (1974) returned to Bealer *et*

OPCS Classification of rural areas

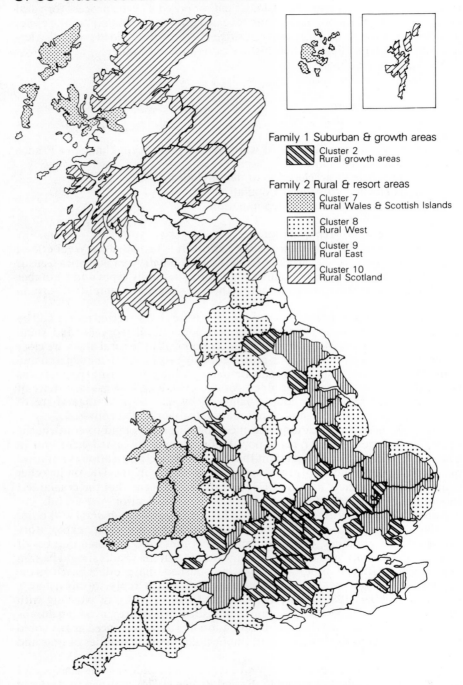

Family 1 Suburban & growth areas
  Cluster 2
  Rural growth areas

Family 2 Rural & resort areas
  Cluster 7
  Rural Wales & Scottish Islands
  Cluster 8
  Rural West
  Cluster 9
  Rural East
  Cluster 10
  Rural Scotland

Fig 5.6   Webber and Craig's classification of rural areas. Source: Webber and Craig (1978)

al.'s (1965) three dimensions and in a survey of heads of household in Illinois, found that socioeconomic status generally accounted for more of the rural–urban variation than either occupation or residence, and then rejected Bealer's approach when they argued that future work should concentrate on single-dimension variables. Nonetheless, most recent work into the psychological attitude of people as to whether areas are urban or rural has used multivariate techniques. For example, Miller and Luloff (1981) used factor analysis to argue that a rural culture still exists in America, and used multiple discriminant analysis to suggest that although residence and occupation are correlated with a rural culture, more central variables could be place of residence at the age of 16, along with several personal demographic features such as religion, income and age. In a British example, Palmer *et al.* (1977) used a semantic differential questionnaire and factor analysis to define the countryside image, and though the findings were difficult to interpret, they suggested that the key variables could be crowding and wilderness, thus returning the argument to one of density.

One other key variable that has traditionally been used to delineate rural areas is the percentages employed in agriculture and forestry. However, although this still may be effective in the less-developed world where about 50 per cent of the world's labour force are still engaged in agriculture (Grigg, 1976) it can no longer be used in the developed world for, as the next section shows, primary employment in most western countries has fallen below 10 or even 5 per cent.

Employment in rural areas

At the outset it is necessary to reiterate the point made at the beginning of this chapter, namely, that accurate up-to-date and spatially relevant data on rural employment, like rural population, are difficult to obtain. There are two main reasons for this. First, the areal units used often contain substantial towns within their borders, and second, the annual and monthly data collected by the Department of Employment only record 'employees in employment'. The following categories of workers are excluded: working proprietors, partners, the self-employed, wives working for husbands, husbands working for wives, persons working in their own homes, former employees on the payroll as pensioners only, and private domestic staff working in private households (McCallum and Adams, 1981). In rural areas these categories are often over-represented. For example, most farmers are self-employed, many farmers now work part-time in other occupations but may still be classified as self-employed farmers, farmers' wives often work for them as partners, professional people may often work from home preferring a rural to an urban base, and large estates still employ a number of private domestic staff.

McCallum and Adams (1981) have pointed out that this can lead to major discrepancies. for example, the Department of Agriculture for Scotland (DAFS) estimated there to be 75,000 people working in agriculture in Scotland in 1976, but the Department of Employment data for the same year revealed only 33,000 people, or just 44 per cent of the DAFS total. In another case, Department of Employment data for the Western Isles of Scotland showed only 26.4 per cent of the population in employment, compared to 42 per cent for Scotland as a whole, and clearly this discrepancy is largely due to the definitional problems outlined above.

In spite of problems with the data, a number of themes and trends do however emerge. The most certain of these is that employment in the primary sector has fallen for over 100 years, and continues to fall by around 2 to 3 per cent a year. Originally most of the loss occurred among farm workers, but now as Chapter 2 demonstrates, farmers themselves are retiring from the industry in increasing numbers, and not being replaced. The reason for the decline is not always poverty (which it often is for smaller farms or those in marginal areas) but in the better areas is due to increasing agricultural efficiency, and even in the mid-1980s to agricultural surpluses (see Chapter 9). For example, Drudy (1978) in a study of a prosperous farming area, north Norfolk, found that changes in agriculture were the foremost reason why the population of the area had decreased by 3.7 per cent between 1951 and 1961, and by 6.9 per cent between 1961 and 1971. (Note however, that Figure 5.2 shows a population increase of between 5 and 14.9 per cent for the same area between 1971 and 1981). In more detail, a study of 153 farms in the area found that redundancy was the most important single reason for leaving farm employment between 1960 and 1970, as shown in Table 5.7.

Nonetheless, as Table 5.8 shows, employment in primary industry can still account for 12 per cent of all employment in rural areas, and when all the ancillary trades asssociated with agriculture are counted in, for example, food processing, machinery, and fertilizer and pesticide production, agriculture may well account for some 10 to 15 per cent of all national employment, although there are no accurate figures for this. What has happened, therefore, is a transfer of production from the field to the factory floor, so that the field is now but one stage of the production-line process.

The decline of agricultural employment can also be related to wider depopulation. For example, Drudy (1978) found evidence that depopulation had been worse in small settlements, and concluded that a low level of service provision, when added to the lack of a sufficient employment alternative to agriculture, could lead to a vicious circle of decline based on the theory of cumulative causation, already outlined. And Hodge and Whitby (1981), the theory's authors, have indeed argued that rural labour markets are at the core of the problem of rural depopulation in developed countries.

In this respect the key variable in the 1980s is the large number of young people entering the rural labour market, following the baby boom of the 1960s. Many studies of previous generations have shown that many young people leave rural areas before they even enter the job market, and that these migrants are the most

Table 5.7: Reasons for mobility from Norfolk farms 1960–70

	Number	%
Redundancy	154	32.3
Low wages	111	23.3
Dissatisfied with work or conditions	86	18.0
Dismissed	35	7.3
Attracted by other employment	24	3.0
Other reasons	67	16.1
Total	477	100.0

Source: Drudy (1978).

Table 5.8: Trends in employment sector for selected counties

Period	County	Primary	Manufacturing	Service	Total
1971–73	Cumbria	− 7.2	+ 2.0	+ 4.5	+ 2.9
1971–75	Lincolnshire	− 1.3	+ 0.7	+ 2.0	+ 1.2
1971–75	Norfolk	− 11.1	+ 2.1	+ 18.3	+ 10.3
1971–75	Cornwall	− 11.2	− 0.3	+ 13.8	+ 7.0
1971–75	Shropshire	− 14.7	+ 3.7	+ 12.3	+ 6.6
1976	Rural areas	12	39	49	100
1976	England	2.5	42	55.5	100

Source: NCVO (1980a).

able and motivated (Rieger, 1972; Hannan, 1969) thus adding weight to the theory of cumulative causation, as the less able young, the old, and the redundant come to form a greater and greater proportion of the population. More recent information on this topic has been provided by Gilg (1980a) in a report on a Ph D thesis by Sally Dench. As Table 5.9 shows, this thesis not only revealed major differences between male and female aspirations, but also major gaps between aspiration and experience. For example, only 11.4 per cent overall favoured factory work, but 21.8 per cent had actually obtained such work, and while 12.5 overall favoured clerical work only 2.6 had obtained such work.

Accordingly, Drudy and Drudy (1979) have advocated the provision of professional and office jobs, and Gilg (1976) has advocated more jobs in office employment and tourism, as well as jobs in manufacturing, as an alternative to agricultural employment, so that not only the aspirations of the whole ability range of young people may be met (Drudy, 1975) but also to increase the level of female activity rates found in many rural areas (Moseley and Darby, 1978).

However, until very recently the accepted wisdom (CRC, 1977a) has been that only manufacturing jobs can provide the sort of economic basis needed for sustained job provision, and that neither hill farming or forestry, as shown in Table 5.10, or service employment and tourism could provide a long-term solution to rural employment problems. However, these analyses have been largely invalidated by the employment 'counterurbanization' of the countryside outlined in the following paragraphs.

Conceptual arguments in favour of employment deconcentration can be traced originally to the need to decant industry from over-congested conurbations to adjacent market towns (Woods, 1968) and then to an acceptance that the extreme population concentrations of older industrial societies was neither economically necessary or inevitable (Commins, 1978).

Empirical evidence also began to build up, and Spooner (1972), in a study of employers who had moved to the southwest found that more of them cited the attractive environment of Devon and Cornwall as reasons for moving to the area, than its labour supply, and Lonsdale and Browning (1971, 267) found that manufacturing plants in 10 southern USA States were more orientated to rural than urban areas and that: 'manufacturing firms appear to be placing increased emphasis on rural and small town sites'.

The main reason for this shift was the reduced importance of the major variable in most economic location models, transport costs, as improved transport and telecommunication facilities have overcome the friction of distance, particularly for the new 'sunrise' industries of micro-electronics and

Table 5.9:　Youth employment, aspiration and experience in mid-Devon

Employment sector	Job preferences of fifth formers		Jobs obtained and desired by a different cohort of 16 year old school leavers	
	Male	Female	Job held	Job desired*
Agriculture and other primary	18.6	11.2	24.3	16.9
Construction	22.0	0.4	16.5	14.7
Factory work	8.3	10.9	21.8	11.4
Sales and distribution	5.4	6.0	9.6	9.8
Clerical	2.1	22.1	2.6	12.5
Services	10.0	19.5	13.0	17.9
Armed Forces	13.7	4.1	8.7	6.5
Professional activities	11.6	18.7	1.8	3.8
Don't know	8.3	7.1	1.7	6.5
Sample size	241	267	115	184

* Cohort of young people still unemployed.
Source: Gilg (1980a).

Table 5.10:　Different costs for providing various rural employment options

£ per job	Public expenditure costs				Net resource benefits effects		
	Light industry	Hill farming	State forestry	Private forestry	Hill farming	Forestry	Light industry
Rural Northumberland	1,300	4,000	29,000	12,500	1,000	−27,000	1,270
Merioneth	1,280	3,700	29,000	8,000	1,100	−23,000	810
Montgomery	1,285	2,200	29,000	8,000	300	−23,000	810
Rural Carmarthen	1,305	4,300	23,000	8,000	1,700	−23,000	820
Sutherland	1,300	3,400	30,000	13,000	300	−23,000	770
Roxburgh	1,350	1,300	32,000	12,500	500	−22,000	300
Wigtown	1,250	7,500	29,000	11,000	2,600	−19,000	1,650
Average	1,300	3,800	28,700	10,700	1,050	−22,800	920

Source: Treasury (1976).

software design. Furthermore, Hodge and Whitby (1982, 26) have pointed out that many employers in small rural businesses are now more able to exploit the advantages of rural locations: 'cheaper space, less congestion, attractive environments and a lower likelihood of industrial disputes'.

The net result is that rural areas have gained an increasing share of manufacturing and service employment as both Table 5.8 and Figure 5.7 show. This is not however mainly due to migration, as in the case of population, or to neo-classical location theories, but according to Fothergill and Gudgin (1979; 1982), the main process differentiating urban and rural areas is no longer structural variation, but appears to arise primarily from indigenous performance, i.e. from different growth among similar firms in different areas, rather than from the movement of firms or changes in industrial structure. In another study Frost and Spence (1984, 145) also argue that industrial restructuring has had a negligible impact on changes in employment patterns, either positively or negatively, and that these are anyway offset by 'very rapid growth in certain service activities', notably the health services, to produce the different patterns of change shown in Figure 5.7. Indeed Frost and Spence conclude that the main result of their detailed study of employment change between 1971 and 1977 must be (1984, 146) 'the critical role that service activities and particularly the widespread public services have played in influencing the nature and patterns of employment change over the period'. Nonetheless, for the rural geographer the most important conclusions of Frost and Spence's work are that:

1   The important and all pervasive theme of decentralization is confirmed;
2   It is a fact that the main growth zones of the country lie in the inner metropolitan periphery; and
3   Core decline and peripheral growth is verified as a principal feature of the nation's changing distribution of employment.

The only drawback to both the studies, by Fothergill and Gudgin, and Frost and Spence, is that they predate the economic recession of the 1980s and the reduction in public sector employment pursued by the Conservative government since 1979. One clue is however provided by Owen *et al.* (1983), who used partial data (unemployment registers classified by last employment) for the period 1979-82 to conclude that the regional north/south split was as important as the urban/rural split, and that future growth would probably be dominated by free-standing towns in the 'sun-belt' of the southeast.

These changes are not confined to the UK and Keeble *et al.* (1983) have found a relative shift of manufacturing industry from highly urbanized to rural regions across the entire European Community, and Bradshaw and Blakeley (1983) have pointed out three significant trends in the US rural economy: first, that agriculture is no longer the dominant employer; second, that manufacturing employment has grown in rural areas; and third, that services have expanded to employ nearly 60 per cent of the rural labour force and to provide the new basic economy for the growing rural population.

Furthermore, these changes have spread to not only the more isolated areas but also to the protected landscapes of national parks. For example, Hedger (1981) in a study of mid-Wales found a massive resurgence in population, housing and employment, but pointed out that manufacturing had only provided 120 new jobs while service employment has provided over 1,000 new jobs between 1971 and 1977. Similar findings have been reported for both mid and east Devon where both service and professional activities have increased considerably (Glyn-Jones, 1979; 1982). Within the national parks 'service industry and in particular tourism is a large and growing sector of the economy of each park' (TRRU, 1982, 127) and this experience has raised doubts about the validity of 'the conventional wisdom that the creation of employment to replace jobs lost from primary industries is best pursued by encouraging small-scale manufacturing industry' (Countryside Commission, 1981a, 73).

These doubts will be considered more fully in Chapter 9 but they also raise the prospect of a radically different future. For looking ahead even a mere 10 years or so, it is possible and even probable, that a much smaller proportion of people's lives will be spent in paid employment (Gilg, 1983c), and that people will retire earlier, work far fewer hours and days per week, and will have very long holidays, so that work can be shared around. The effect on population and employment patterns could be profound and Blake (1979, 10) has suggested that:

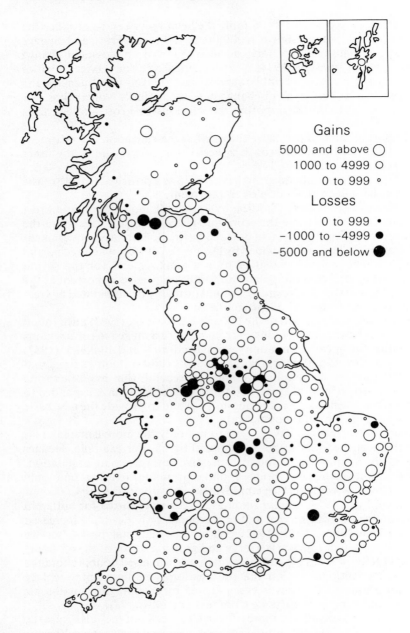

Total employment change, 1971–77 (absolute)

Fig 5.7   Absolute and relative employment change 1971–77. Source: Frost and Spence (1984)

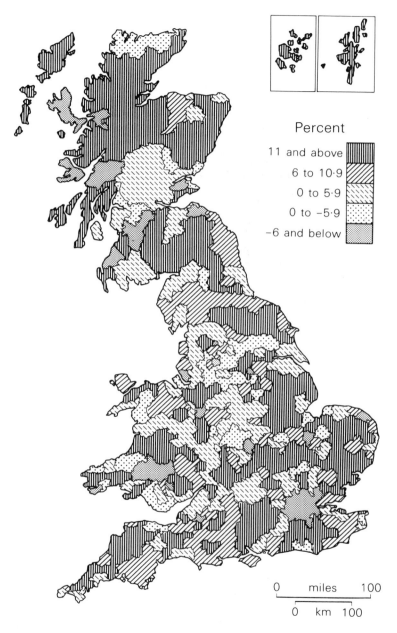

Percent

| 11 and above |
| 6 to 10·9 |
| 0 to 5·9 |
| 0 to −5·9 |
| −6 and below |

Total employment change, 1971–77 (percent)

The constrictions of our older cities will become more and more irksome. There will be a demand for more second homes in the countryside, and some families may reverse the process altogether by moving to the country and keeping a town pad for use while at work.

In other scenarios, almost everyone could move to low-density countryside houses and communicate via the electronic technology that is already available, or alternatively people could recolonize the countryside and run smallholdings (Anderson, 1975). If these forecasts are correct, and they of course ignore the inertia of the system, and the immense social capital invested in cities, then the population and employment changes revealed by this chapter are only the precursors of a massive recolonization of the countryside, not only in the UK but also the USA (Fisher and Mitchelson, 1981). However, before this arcadian millennium is reached, the real rural problems of today still have to be resolved, namely poor rural transport and accessibility, low levels of service provision and rural deprivation. Accordingly these are considered in the next chapter.

6

Rural transport, service provision and deprivation

Change in rural transport is not only the key to many of the changes outlined in Chapters 4 and 5, but it is also a central connecting factor linking service provision and rural deprivation. At the outset however, it should be stressed that many of the changes have been beneficial. In particular, the massive increase in car ownership from about 10 per cent in the early 1950s, to about 58 per cent of all urban households and around 70 per cent of rural households in 1982 (Phillips and Williams, 1984) has let more and more rural people travel to a greater range of destinations at their own time and convenience. At the same time, it has allowed urban-based employees to commute from not only exurban areas (Lawton, 1968) but also remote rural areas (Lewis, 1967) in ever increasing numbers and these commuting fields have been used as another, albeit disputed way of defining rural areas (Ball, 1980). In addition, increasing car ownership and high-speed road provision has been a main factor in the massive increase of rural recreation enjoyed by many people since the 1950s, and discussed in Chapter 7.

Nonetheless, although for many people the rise in car ownership can be seen as a great success, for a significant minority either unable to drive or unable to afford a car, notably the old, the young, the unemployed, the housewife (left at home without a car) and the disabled the corollary of increased car ownership has been a rapid decline in rural public transport, a concentration of services into ever more remote central places with increased journey lengths as shown in Table 6.1a, and a real sense of isolation and deprivation. Ironically, the advent of modern telecommunications has often increased the sense of physical isolation, by demonstrating more vividly the opportunities being missed and by heightening (Clark and Unwin, 1981, 55) 'people's awareness of their spatial immobility'. Most recent geographical research work has concentrated on the three related problems of declining public transport, centralization and reduction of rural services, and increased rural deprivation. Accordingly, the rest of this chapter discusses these three topics in turn.

Rural transport and accessibility

Although some studies have been made of rural railways, the majority have looked at road-based public transport (Halsall and Turton, 1979), since only the road provides the flexibility required in rural areas. Until recent innovations, that meant the rural bus. As Table 6.1b shows, however, bus journeys have been in rapid decline, and as a consequence 'many rural bus services have been reduced severely in frequency and part routes and whole routes have been withdrawn' (Knowles, 1978, 668). This reduction would have been even more

Table 6.1: Some features of rural transport

A   *Modal split and journey lengths in rural areas*

Purpose	Modal split*	Journey lengths in miles Rural area	Mean for all areas	% difference
Work	29	7.91	5.75	+36
Education	23	4.10	2.56	+60
Shopping	36	6.24	3.44	+81
Social	10	9.95	7.40	+34
Escort	0	7.79	4.88	+60
Other	2	15.60	10.94	+43
All purposes		8.85	6.30	+41

B   *Modal split in rural areas*

Mode	Distance travelled — percentage 1965	1976/75
Car or van as driver	39	48
Car or van as passenger	32	33
Motorcycle or bicycle	3	2
Walk (over 1.6 km)	2	1
Local bus	13	5
Long distance bus, works or school bus, train etc.	11	10

* Data from a survey in Herefordshire. Other data from National Travel Survey.
Sources: Banister (1983); HC (1978).

severe if subsidies hadn't been paid since 1968 and some of the innovations shown in Table 6.2 hadn't been introduced.

Nonetheless, a 1978 survey of innovative rural bus services found that only one was making a profit, and most were making losses of between £14 and £48 a week, albeit smaller losses than the £64 to £347 a week losses experienced on some conventional routes (HC Select Committee, 1978). The main reason for the continual losses, apart from the switch to private car transport already outlined, is that the modal split outlined in Table 6.1a makes it very difficult to provide an economically efficient service. For example, in the Herefordshire survey shown in Table 6.1a the 36 per cent of trips accounted for by shopping involved about 700 passengers per hour between 10 a.m. and 5 p.m.; the 29 per cent of trips accounted for by work involved about 1,000 passengers per hour between 7 a.m. and 9a.m. and around 500 passengers per hour between 4 p.m. and 7 p.m.; and the 23 per cent of trips accounted for by education involved about 2,000 passengers per hour between 8 a.m. and 9 a.m. and around 1,000 passengers per hour between 3 p.m. and 5 p.m. It is clearly very difficult to provide an economic service to provide for such peaks and troughs in demands, and so most attempts to ameliorate the decline in rural public transport have tried to be 'user-specific' or to use 'spare capacity' on services already in existence.

Examples of the 'user-specific' approach are provided by commercial minibuses and midibuses, hired village buses, community buses, and social car schemes. Examples of the 'spare capacity' approach are provided by post buses and school buses (National Consumer Council, 1978). However, although the National Consumer Council concluded that such schemes do make a very real contribution, and also make possible extra journeys and excursions which would not be provided by fully routed bus services, they also pointed out that the schemes could not provide a complete answer to accessibility problems in

Table 6.2: Principal transport policy changes and their impact on rural areas

	Date	Recommendations and actions
Jack Committee's Report on Rural Bus Services	1961	Selective direct financial assistance to unremunerative services. Fuel tax rebate.
Beeching Report on Railways	1963	Closure of 5,000 km of rural railways.
Series of surveys in six areas	1963	To examine the problem of and the demand for rural transport services.
Transport Act	1968	System of direct grants with central government paying half where the services covered half of their operating costs. Revenue costs to cover losses on unremunerative rail services. National Bus Company set up. School service contracts could now take fare-paying passengers if there was excess capacity. Fuel-tax rebates increased. New bus grants introduced. Concessionary fares.
Local Government Act	1972	Coordinating function for the County Councils through Transport Policies and Programmes (TPPs). County Councils to administer the distribution of the Transport Supplementary Grant.
Passenger Vehicle (Experimental Areas) Act	1977	Relaxation of the licensing laws in certain areas so that innovative services could be introduced.
	1977	Rural Transport Experiments (RUTEX) set up in four 'deep' rural areas.
Transport Act	1978	County Public Transport Plans to coordinate passenger transport to meet the 'needs' of the public. Guidelines set for concessionary fares. Minibuses (8–16) seats exempt from Public Service Vehicle Licences and Road Service Licences, provided that the drivers were from approved voluntary organizations. Car sharing allowed for payment.
	1978	Market Analysis Projects (MAP) to match service provision more closely with expressed demand.
Transport Act	1980	Major change in bus licensing — deregulation. Small vehicles (fewer than 8 seats) no longer classified as Public Service Vehicles. Express services (over 30 miles) no longer require Road Service Licences. Road Service Licences 'create a presumption in favour of the applicant'. Trial Areas can be designated where there are no Road Service Licences.
White Paper Proposals	1984	End of road service licensing. Break up and privatization of National Bus Company. Greater encouragement of wider use of services run by education, health and social service authorities, the Post Office and others.

Sources: Banister (1983); Transport (1984).

rural areas. Furthermore, they appeared to work best in areas of good conventional provision, where they could plug the gaps, but worked least well in areas of poor conventional provision.

A weakness of both the conventional and innovative approaches, however, is that they have been dominated by not only the need to break even financially, but also by the 'behavioural approach' to and the financial view of the problem, which collectively (Stanley and Farrington, 1981, 65) have:

> engendered studies that try to enumerate the demands for transport in rural areas and then attempt to change network and service configuration to minimize the subsidy needed and meet the demand so enumerated. Demand is normally measured by collecting information about the behaviour pattern of individuals, and finding out how and why people are travelling. The dominance of the behavioural approach has led to the equating of need with willingness to pay or expressed demand . . . [but] a by-product of behaviour-based transport planning is that individuals who do not show 'demand' for movements are assumed not to 'need' to move

Therefore one major trend in rural public transport studies has been a move away from 'behavioural' to 'need' studies and in some cases a joint approach. For example, Peat, Marwick and Mitchell (1980) from a study of 'need' in Dyfed in Wales, not only produced the stratified definition of 'need' shown in Table 6.3 but from a 'behavioural' study also produced the fourfold stratification of need shown below:

Essential
First priority   Household shopping, chemist
Second priority   Health services, bill payments, repair services, Post Office and other personal business

Desirable
Third priority   Social visits (including hospital), public meetings, school activities, gift shopping
Fourth priority   Outings and special events, sports activities, social activities

Implicit in both stratified approaches adopted by Peat, Marwick, Mitchell & Co. is the concept of time–space geography shown in Figure 6.1. This approach has been used by Moseley (1979) to measure accessibility by constructing a time–space matrix for each rural/social/age group (rows) and for each rural service needed (columns). Other methods to measure accessibility have involved questionnaire surveys of hospital visits (Haynes and Bentham, 1979) or matrices of distance from various centres of population as shown in Table 6.4, or surveys of access to a car as shown in Table 6.5.

At its most basic level then, accessibility (Moseley, 1979, 7) measures three components:

1 people, the residents of rural areas;
2 the activities or services which they require;
3 the transport or communication links between the two.

A number of authors have tried to produce statistical models for numerically evaluating accessibility. For example, Robertson (1976) used real road distance and estimated population for 5 km squares in Argyll to test an algorithm in which services search over the population surface until locations are found which minimize the cost, time or effort involved in people travelling to these services. In another study in Wales, Nutley (1980a) used around 10 indices, including bus services, distance to shopping centres, and access to Cardiff, to

Table 6.3: Overall definition of public transport needs in rural areas

Element of need	Client population	Access requirement	Required frequency
Primary needs			
School transport	School children (over 8 and living more than 3 miles from school or under 8 and living more than 2 miles from school)	Primary or Secondary school	School days
Shopping transport	Persons in non-car-owning households	Shopping centre	Weekly
Health transport	Persons in non-car-owning households	Doctor's surgery	Weekly
Secondary needs			
Work transport	Employed persons in non-car-owning household	Employment centre	Weekdays
School transport	School children (over 8 and living between 2 and 3 miles from school or under 8 and living between 1 or 2 miles from school)	Primary or Secondary school	School days
	Secondary school children	Secondary school (after hours)	Weekly
Transport to Further Education	Persons aged between 16 and 18 who have left school	College of Further Education	Weekdays during term
Shopping transport	Housewives in households where the car is absent during the day	Shopping centre	Weekly
Social transport	Persons aged 70 or over in non-car-owning households	Day centre	Weekly
	Persons in non-car-owning households	Hospital visiting	Weekly

Source: Peat, Marwick, Mitchell (1980).

produce a composite index, and in a Scottish example Nutley (1979, 153) has also argued that a number of a different methods should always be used, 'as no single method can adquately represent the various alternative conceptions' of accessibility. Accordingly Nutley used three measures of accessibility in a study of the Highlands and Islands of Scotland, namely travel timings from a central point, shortest paths through a network between every pair of nodes, and potential surfaces; and in yet another study of Scotland, Nutley (1980b) used multiple regression to test both objective (i.e. bus capacity) and subjective (i.e. people's perceptions) measures of accessibility.

All of these different methods however produce significant measures of mobility deprivation in rural areas, and since neither the conventional nor innovative approaches to public transport have been shown to do more than ameliorate or delay the decline in rural transport provision, many authors have called either for some of the options in Figure 6.1b or for a 'total approach' to the problem. For example, Moseley *et al.* (1977) have argued that resources should be shifted between spending heads and even between agencies, and Banister (1980) has argued for a total welfare perspective which involves the best use of all existing facilities, not just the transport-related ones. Finally, Stanley and Farrington (1981, 78) also conclude that public transport is only one

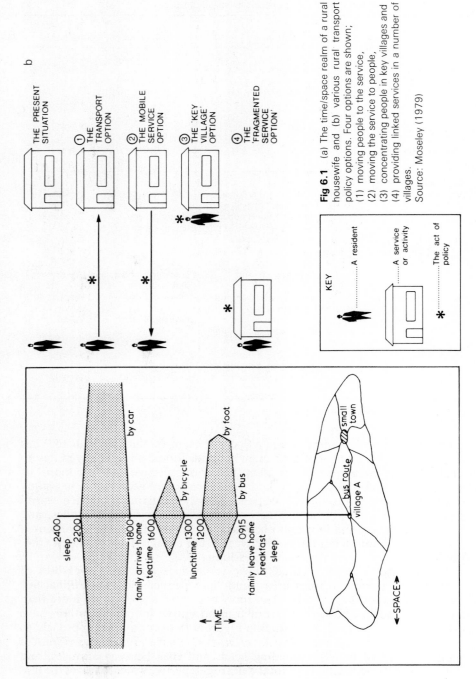

Fig 6.1 (a) The time/space realm of a rural housewife and (b) various rural transport policy options. Four options are shown;

(1) moving people to the service,
(2) moving the service to people,
(3) concentrating people in key villages and
(4) providing linked services in a number of villages.

Source: Moseley (1979)

Table 6.4: Number of rural settlements by size and distance from an urban centre

| Population size group | Distance in miles from an urban centre with over 20,000 population | | | |
	5–9.9	10–19.9	20 +	Total
5,000–19,999	140	80	30	250
2,000–4,999	270	150	70	490
1,000–1,999	420	260	110	790
500–999	690	480	210	1,380
0–499	2,540	2,400	850	5,790
Total	4,060	3,370	1,270	8,700

Source: Norfolk (1979).

Table 6.5: Access to a car by social/age groups (per cent)

	Employed persons	Housewives	Retired persons	All adults
No access at all	5.6	12.3	40.3	16.3
No access during the day	15.8	42.4	10.9	20.0
Regular access	78.6	45.3	48.8	63.7

Source: Peat, Marwick, Mitchell (1980).

component of accessibility and argue that accessibility deprivation can be alleviated in one of three strategic ways:

1 By improving the socioeconomic conditions of the population;
2 by redesigning the provision of facilities required by the population; and
3 by providing public transport explicity tailored to the population's needs.

Two of these strategies directly relate to the topics to be considered next in this chapter, namely (2) the provision of facilities (rural service provision) and (1) socioeconomic conditions (rural deprivation) and so attention is now turned to rural service provision and rural deprivation.

Rural service provision

Almost any collection of papers on rural services, even those conceived and executed in virtual independence, will produce two clear themes (Moseley, 1978), first, an inadequate access to opportunities and rural services, as shown in Table 6.6, and second, a feeling among rural residents that their needs are not being adequately reflected in the policies and services produced by local and central government agencies (Smart and Wright, 1983). A third and almost as clear a theme is that even this inadequate level of services has been declining very fast, in both the USA (Johansen and Fuguitt, 1984) and in the UK. For example, 13 per cent of villages in Gloucestershire and Wiltshire lost their village shop between 1972 and 1977 (Standing Conference of RCCs, 1978) and the number of rural bus services in north Norfolk fell by 17 per cent between 1975 and 1981 (Harman, 1982). There are two clear reasons for these losses (Woollett, 1981). First, both private and public services have raised their population thresholds (Bell *et al.* 1974) in a desire to achieve the economies of scale concentration (already discussed in Chapter 4), and second, the dramatic increase in

Table 6.6: Inadequate levels of rural service provision

A   Service facilities by villages in West Dorset (% with facility)

Village size	No.	Village hall	Pub	Garage	Post office	Shop	School
0–100	30	24	7	0	14	0	0
101–500	34	94	69	60	83	77	49
500–2,000	10	90	90	90	100	100	90
2,000+	1	100	100	100	100	100	100

B   Estimated total rural population lacking selected services

Population + (millions)	Service	Population* (millions)	Service
6.1	Dispensing chemist	2.2	Social Security office
4.9	Infant welfare clinic	1.7	Citizens' Advice Bureau
3.6	Doctor's surgery	1.0	Social Services office
1.4	Village hall	0.3	Job centre
0.7	Post office/Food shop	0.1	Permanent library

+ Population of parishes lacking service.
* Population over 10 miles from service.
Sources: Standing Conference (1978) and Norfolk (1979).

accessibility for those with private cars has allowed the majority of people to travel much further distances to better and/or cheaper facilities (Rowley, 1971). Both these reasons have combined to produce a vicious circle of decline.

Virtually no service has been immune from this process. The two most likely services to trigger a process of decline are the post office, which often acts as a multiple service facility (Taylor and Emerson, 1981), and the primary school, which is not only a central feature of the community but also a vital ingredient in retaining a young and active population (Jones, 1980). If either or both these facilities close, the village shop may well be left to cater for only the poor, the old and the immobile, and since the shop on its low turnover has to charge high prices, these people are doubly disadvantaged (Harman, 1978). One final nail in the coffin may be the accelerating rate of rural petrol station closures since this would further encourage the car owner to leave the local village for petrol and other services (Dean, 1983). In summary, the decline and concentration of rural services has hit the most deprived sections of rural society hardest, and nowhere is this more so than in health care, where the closure of rural health facilities has meant that those in greatest need, women, the old, and the poor, now have least access to the services they need (Haynes and Bentham, 1979).

The picture of declining services presented above does, however, tend to ignore the impact of the counterurbanization processes discussed in Chapter 4. There is little evidence so far, though, to suggest that counterurbanization has halted the decline of service provision, since population thresholds for services have at the same time been rising (Johansen and Fuguitt, 1984). What evidence there is from the USA suggests that local authorities there are simply not prepared, either physically or mentally, to deal with the new service demands of the exurban inmigrants (Green, 1983) although work in Canada, using a game-playing approach, has tended to suggest that the inmigrants have the same service demands as the existing inhabitants (Joseph and Smit, 1983).

The economics of service provision, and the arguments in favour of service concentration, especially when savings can be made across all the services (Warford, 1969) have already been discussed in Chapter 4. Although much recent work (Maos, 1983, 47) confirms that 'the efficiency of services is higher in concentrated settlement patterns and can be further improved by the transfer of service functions from lower to higher ranking centres', some recent work has called for less emphasis to be placed on the economic evaluation of service provision. For example, Askew (1983) has produced a welfare model which attempts to resolve the efficiency–equity conflict inherent in public facility location modelling. Woollett (1981) has shown not only how rural people can provide their own alternative services but can also further extend them, and Clark (1980) has argued that if we want to sustain rural services it will have to be by self-help. Nonetheless, it is often the most deprived areas or people who are least able to help themselves, and in many rural areas the decline in rural transport and rural services outlined above has led to real rural deprivation.

Rural deprivation

Unlike urban deprivation,which is concentrated in large areas of run-down property and visibly apparent physical decay, rural deprivation is spread more thinly. This can lead (Walker, 1978, 3) to the 'ecological fallacy' in which theories of deprivation use average scores of deprivation measures to pin-point deprivation 'black spots', which under this approach are far more likely to be found in urban rather than rural areas. Nonetheless as Moseley (1980) has pointed out, there are many areas of overlap between urban and rural deprivation, particularly in the inner urban and outer rural areas, as Figure 6.2 shows. Moreover, in the USA the 11 million rural people with incomes below the poverty threshold account for a disproportionately large share of the nation's poor, 38 per cent (Daft, 1983, 75) and 'one can find small pockets of rural poverty spread throughout almost every part of the nation'.

However, most workers stress that rural deprivation isn't a function of rurality but is instead a product of the national social structure which according to Neate (1981, 20–1) is then:

> compounded by the problems of poor accessibility in areas that are relatively (or absolutely) sparsely populated and increasingly deprived by the concentration of employment and services in centres of population.

And Walker (1978) has argued that although all rural dwellers are burdened by the extra costs of gaining access to services, doing without them, or consuming inadequate services, that these costs bear disproportionately on the rural poor.

Within the overall concept of deprivation, a rural deprivation cycle has been identified in which three types of deprivation (Shaw, 1979; ACC, 1979) interact as shown below:

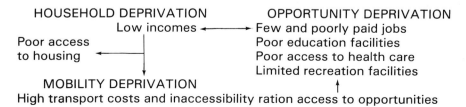

HOUSEHOLD DEPRIVATION   OPPORTUNITY DEPRIVATION
      Low incomes ◄──────► Few and poorly paid jobs
Poor access   │   Poor education facilities
to housing ◄─────   Poor access to health care
           ▼   Limited recreation facilities
    MOBILITY DEPRIVATION   ↑
High transport costs and inaccessibility ration access to opportunities

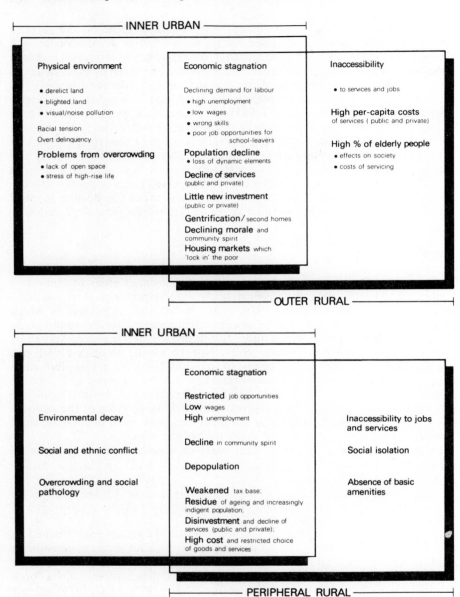

Fig 6.2 Two related models of the relationship between urban and rural deprivation. Sources: Moseley (1980) top diagram and Pacione (1984) bottom diagram.

Mclaughlin (1981) has added one further item to these factors, namely the problems relating to the dominance of the prevailing Conservative political ideology in rural areas, which emphasizes enterprise and the market at the expense of public sector responsibility and support (Newby *et al.*, 1978). Newby (1977, 83) has also shown how this same group, the farmers, have managed to keep farm workers deprived, partly because the conditions of the recent past were so bad with:

> near starvation wages, wretched housing, benevolent paternalism or arbitrarian despotism and anti-trade-union oppression . . . that it now . . . seems difficult to believe that, despite the amelioration of his circumstances, in relative terms the position of the agricultural worker has remained substantially the same.

Newby therefore introduces one further element into the rural deprivation model, namely the relativity of deprivation, and reasserts the two basic problems of the model, the apparent well-being of rural areas based on their attractive environment, and the difficulty of defining deprivation when it is spread in small pockets over much of the countryside.

Accordingly, recent work has looked at new ways of defining rural deprivation over and above the arithmetic of woe shown in Tables 6.5 and 6.6. For example, Knox and Cottam (1981a), although they used Scottish parish and district data to confirm Moseley's hypothesis (Figure 6.2) that the most deprived areas are the inner urban and outer rural (when they found that central Glasgow and the Western Isles of Scotland were the most deprived), also argued that better results were obtainable from not only a welfare, but a questionnaire-based approach to individuals' satisfaction or dissatisfaction with their way of life (Knox and Cottam, 1981b). In another example Cullingford and Openshaw (1982) list five different approaches to the identification of rural deprivation: income, individual perceptions of quality of life, accessibility, distribution of facilities, and social area analysis by multivariate statistics. Although social area analysis suffers from the 'ecological fallacy' already outlined, Cullingford and Openshaw improved its precision by eliminating urban areas and variables from a cluster analysis of rural deprivation. However, though the only currently practicable approach to identifying rural deprivation was an area based one, and although it has been possible to make some sense of census data for rural areas, they still concluded, in line with Knox and Cottam, that it would be far better if the analysis could be performed at the individual level.

It is clear from the above commentary that a number of rural geographers have become deeply interested in rural deprivation and Cloke and Park (1980, 60) have even suggested that:

> The primary aim within rural geography could become one of ensuring equity of opportunity, well-being and quality of life between all sectors of the rural community, coupled with the creation and maintenance of a productive and self-perpetuating rural environment.

Laudable as these aims are however, they ignore the situation set out at the beginning of this chapter, namely that the majority of rural inhabitants are very satisfied and happy with their life style, and have consciously chosen to live in the countryside. Many of these people are moreover less concerned with the socioeconomic issues discussed in the last three chapters, but more with the changing environment of the countryside and the increasing pressures being placed upon it. Accordingly attention is now turned towards environmental issues in the countryside, beginning with recreation.

7

Rural recreation and tourism

In a recent review of rural leisure and recreation research Owens (1984) has commented on not only how participation in rural leisure and recreation grew rapidly in the 1950s and 1960s, but how it was also accompanied by a surge of interest in applied research which has since spawned an enormous literature. Although this literature has now grown to gargantuan proportions, Owens has also managed to discern a number of themes and trends.

Initially, the subject was dominated by fact finding at the level of both national and regional demand surveys (what people did or wanted to do) and site surveys (what they did when they got there). Until the mid-1970s these types of study dominated the field (Coppock and Duffield, 1975) and produced a very useful baseline for further work. However, these studies also recognized that the strong links and relationships they described (e.g. increasing rates of participation with car ownership) were permissive rather than causal. Accordingly, from the mid-1970s, research began to examine the perception and behaviour of individuals, and then to develop general conceptual and theoretical research, in a shift from the 'inductive' to the 'scientific' route shown in Figure 1.1. Smith (1983) in another recent summary broadly agrees with Owens, and divides his textbook *Recreation Geography* into a four-stage continuum from description, to explanation, to prediction and finally to policy formulation. Accordingly this chapter examines rural recreation from the viewpoint of *recent trends* (broad patterns of recreational use), *impact studies* (more detailed studies of recreational behaviour and its effects on the environment), *second homes* (a specific type of impact study), and *theoretical and quantitative approaches*, before asking whether recreational geography has come of age.

Recent trends

Virtually all commentaries on rural recreation (Blacksell, 1983; Countryside Comission, 1980) point out that the three main factors that explain rates of rural recreation, time, income and mobility, have beneficially increased since the war. Furthermore, the government (Minister of Land, 1966, 3) has not only sanctioned this growth when, as long ago as 1966, they stated 'that townspeople ought to be able to spend their leisure in the countryside if they want to' but following a report of the House of Lords (HL Select Committee 1973) has also endorsed and encouraged recreational uses when they accepted (Secretary of State for the Environment, 1975, 1) that 'recreation should be regarded as one of the community's everyday needs and that provision for it is part of the social services'.

Nonetheless, virtually all studies also conclude that rural recreation rates are still strongly influenced by social class and income as shown in Figure 7.1 and that the more affluent have much greater participation rates as shown in Table 7.1. Other factors not shown in Table 7.1 are education (with higher rates for the better educated) and sex (with lower rates for women (Birch, 1979). Table 7.1 also shows that participation rates are fairly sensitive to economic growth and although 'positive factors' increased between 1977 and 1980, 'negative factors' also increased, to contribute to the marked fall in recreational activity also shown between 1977 and 1980.

Another change has been noted in longer-term recreation, i.e. tourism, which in some rural areas, notably National Parks, is the mainstay of the rural economy (TRRU, 1981). In another area very dependent on tourism, the south-west of England, the eight million tourists who visited the area in 1973 spent £227 million and employed at the peak of the season 100,000 people (South West Economic Planning Council, 1976). However, against this must be set an extra local authority expenditure on services of £4.6 million, and a regional multiplier of only between 0.35 and 0.45, to provide about 5 per cent of regional income. However, not only is domestic tourism static, but the regional multiplier is also falling, since the one major trend in tourism is a move away from high-multiplier tourism in hotels to low-multiplier tourism in self-catering accommodation. Between 1970 and 1982 the percentage of holidaymakers staying in hotels and boarding houses fell from 31 to 27 per cent, while rented accommodation and camping grew from 29 to 35 per cent (BTA, 1983). The trends outlined

Table 7.1: Changes in factors and trends in rural recreation

	1970	1977	1980
Positive factors			
Real disposable income [+]	104	120	140
Annual paid holiday[x]	10	30	75
Number of cars (millions)	10	13	14
Negative factors			
Unemployment [+]	100	260	310
Real price of Petrol* (pence)	160[++]	105	135
Trends			
Social Class AB [+++]	—	67	58
Social Class C1 [+++]	—	55	50
Social Class C2 [+++]	—	55	39
Social Class DE [+++]	—	38	29
Age group 16–19 [+++]	—	59	49
Age group 30–34 [+++]	—	61	50
Age group 60–64 [+++]	—	49	33
Low Income [+++]	—	43	32
Middle Income [+++]	—	58	42
High Income [+++]	—	63	54
Drive, outings[xx]	—	34	26
Long walks[xx]	—	21	17
O trips [+++]	—	46	58
1–8 trips [+++]	—	46	33
Trip distance (kms)	—	53.2	50.1

Source: Countryside Commission (1982).
[+] 1968 Base = 100, [x] % with more than 4 weeks, * At 1980 prices, [++] 1973, [+++] % participation in last 4 weeks, [xx] % in average summer month.

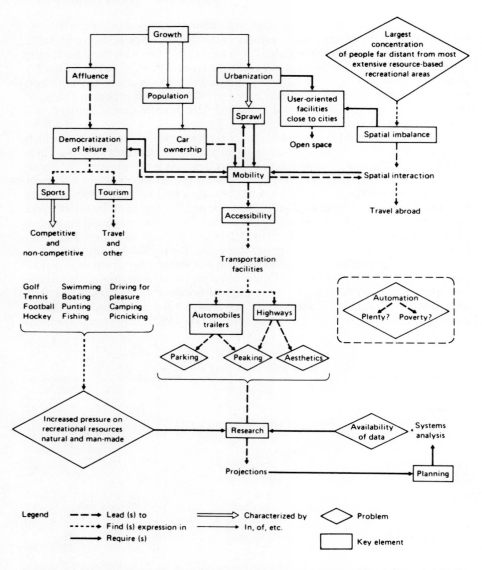

Fig 7.1   Key elements in the demand for outdoor recreation (p. 110) and the decision process in outdoor recreation (p. 111). Sources: Smith (1983) (p.110) and Pigram (1983) (p. 111).

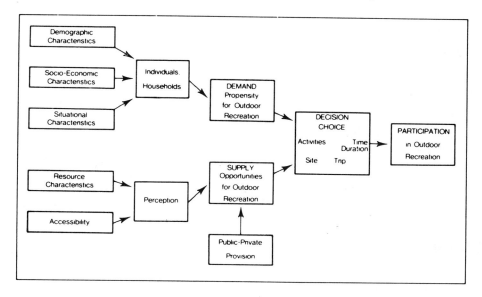

above are, of course, only broad averages to set the scene and so attention is now turned to more detailed and more specifically geographical studies.

Impact studies

Data of the type outlined in Table 7.1 have two drawbacks for the rural geographer. First, they by and large provide only aspatial or coarsely aggregated data, and second, the sample sizes needed to produce statistically meaningful data-sets mean that only national survey organizations can undertake the large surveys needed. However, a good number of geographers have conducted recreational surveys in the last 20 years and a good deal of experience in the methods and techniques for conducting visitor surveys, in particular, has been obtained, so that it is now generally agreed that the following procedure should be followed:

1 Select sample unit and size;
2 Collect information by (a) mechanical techniques or (b) observation techniques or (c) questionnaire techniques; or by a combination of (a), (b) and (c); and
3 Codify and computerize data and finally analyse it (TRRU, 1983).

Since it is easier to achieve a meaningful sample size at the recreation site, most rural geographers have concentrated on these types of study, rather than demand-based questionnaires in the towns and cities. Exceptions are provided by contract work, for example, the Tourism and Recreation Research Unit (TRRU) (1975 to 1977) was sponsored by a number of government agencies to produce four reports on rural recreation in Scotland, based on a survey of 11,600 tourists in their cars and 6,800 Scottish residents in their homes, and in another sponsored survey of 35,810 cars leaving Edinburgh they were able to produce the pattern of recreation shown in Figure 7.2a.

Most work has however concentrated on the recreation site itself and parti-cularly sites in the *uplands*, the *urban fringe* and *wildlife sites*. In the *uplands*, where access is almost universally by car or coach trip, virtually all surveys point out that well over half of visitors never leave either their cars or the car park and that there is a marked and almost complete segregation of the active walker from the passive pursuits of picnicking and scenic viewing (Haffey, 1979). Moreover, as Figure 7.2b shows, there is a marked tendency to congregate at a very few 'honeypot' sites. Accordingly these areas are heavily used and subject to severe environment erosion. Even when people do go for walks, they tend to keep to public paths and these are often also severely eroded (Coleman, 1981). However, in a study of recreational impacts on other land uses, Gibbs (1976) found that only farming was particularly affected, notably by damage to walls and fences and by trespass nuisance, but that the impacts on grouse numbers and water gathering were negligible. Nonetheless, all three uses experienced a greater need to warden or patrol their areas in seasons of recreational use.

In the *urban fringe* it has already been shown in Chapter 3 that recreation can have an adverse impact on farming, but other work has shown that the 500 or so informal recreation sites which cover 5.7 per cent of London's Green Belt (Ferguson and Munton, 1979) are not heavily used by either car-owing suburban dwellers (who leapfrog over the green belt into so-called proper countryside since they perceive the sites as being too near to be worthy of a special car trip) or by carless inner city residents (who spurn the use of public transport) (Harrison, 1983). Instead, the sites are used by local residents, of who some 25 per cent or so come by foot and use the sites like an urban park, primarily to take a walk (Harrison, 1981). Therefore, in an unexpected contrast with the uplands, informal recreation sites in the urban fringe do not normally suffer from recrea-tional over-use.

In wildlife sites however, even a minor increase in visitor use can disturb a habitat and reduce the wildlife interest of the area. A major problem is that many people visit a wildlife site, not because it is one, but because it provides access and a car park. For example, on the north Norfolk coast McLaughlin and Singleton (1979) found that only 15 per cent of visitors had come to birdwatch, compared to 27 and 19 per cent respectively who had come to walk or sunbathe. Not surprisingly, many visitors to the area could also see no problem in picking wild flowers, or allowing dogs to run freely. These activities and the sheer weight of numbers have caused an ever-present danger of a 'blow-out' of the sand dunes, and a subsequent inundation of the sea onto reclaimed farmland behind the dunes. Such major physical impacts are fairly easy to assess, but not enough work has yet been done to assess the effect of visitors on wildlife itself, though it is known that shy species will decline, while common species will often increase, and that fauna both on land and in water will decline faster than flora (Liddle and Scorgie, 1980).

Nonetheless, it has been claimed that the problems of recreation have been overstated and Fitton (1979, 57) has argued that: 'There has been a tendency to view recreation in the countryside as a problem rather than a welcome oppor-tunity for people to enjoy themselves'. One reason for this problem-dominated view is the over-concentration of people into a few undesigned places, largely because people do not have either the knowledge or the confidence to leave the beaten track, whether it be road or footpath. There is however much evidence that people want to be more adventurous, and to learn more about the country-

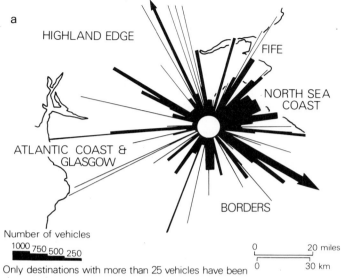

a

HIGHLAND EDGE

FIFE

NORTH SEA COAST

ATLANTIC COAST & GLASGOW

BORDERS

Number of vehicles

1000 750 500 250

0   20 miles

0   30 km

Only destinations with more than 25 vehicles have been shown. Journeys within survey cordons

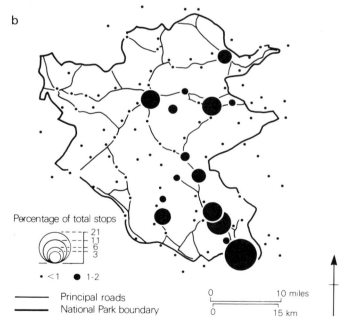

b

Percentage of total stops

21
11
6
3

• <1   ● 1-2

——— Principal roads
■■■ National Park boundary

0   10 miles

0   15 km

Fig 7.2 Examples of origin–destination and recreation site surveys. In the upper diagram (a) a cordon survey of motorists leaving Edinburgh produced the marked coastward-bound pattern shown, while in the lower diagram (b) a survey of visitors at each site produced the pattern of concentrated stops shown in the Yorkshire Dales National Park. Sources: Coppock and Duffield (1975) upper and Patmore (1983) lower diagram.

side through interpretation (Countryside Commissions, 1975) and one excellent way of combining all these attributes is in a specially designed recreational site, the Country Park.

Most of these were set up in the early 1970s, often from existing sites rather than *de novo*. Many of the parks were established near centres of population following the Countryside Act 1968 which provided grant aid for parks in an attempt to relieve congestion on the roads, to ease pressure on more remote and solitary places, and to reduce the risk of damage to vulnerable and valuable areas of countryside (Brotherton, 1975). Although most country parks are monitored by crude counts of visitors, there have been few studies of whether they have achieved the aims set out above. One example has however been provided by Cloke and Park (1982) in their study of Craig-y-Nos Country Park in the Brecon Beacons National Park of South Wales shown in Table 7.2. They concluded that the Country Park had been successful in absorbing some potential visitors from open roadside sites (e.g. the Trecastle Road site), and that many Country Park users were former users of the Trecastle Road site. They also concluded, however, that a significant number of visitors preferred the freedom of the Trecastle Road type of site (an unfenced road running through moorland) to the more organized nature of the Country Park.

In conclusion, all impact studies have confirmed that there is a very strong pattern and indeed concentration of recreational activity along the lines shown in Table 7.3. There are however two further types of recreational impact, both of a longer-term nature, namely *tourism* and *second homes*.

In the case of *tourism*, the economic impact in holiday areas like the southwest of England and the National Parks has already been pointed out. In recent years tourism has also made an increasing impact on farming, particularly in the more scenically attractive and marginal farming areas, and in Scotland, at least 20 per cent of farms indulge in farm tourism, and farm tourism accounts for at least 10 per cent of net income (Denman, 1978). Similar findings have been reported by Davies (1983) for the 'less favoured areas' (the uplands) of England and Wales. In these areas 20 per cent of the farms sampled indulged in farm tourism, but most farmers saw it as a supplement rather than as an alternative to farm income, although Davies argued that it could be a much

Table 7.2: Visitors' attitudes to two types of recreational site

% response	Trecastle Road	Craig-y-Nos Country Park	% response	Trecastle Road	Craig-y-Nos Country Park
Activities			*Most attractive features*		
Nothing	—	1	Natural, peaceful		
Picnic/short walk	39	76	environment	65	57
Walking	16	—	Good walking country	3	8
Natural history	—	4	Water	32	14
Swimming	10	—	Toilets, parking, etc.	—	17
Fishing	7	—			
Multiple	28	19	*Which type of site preferred*		
Too many visitors			Roadside (informal)	74	14
Yes	45	11	Country Park (organized)	16	75
No	52	89	Both equal	10	11

Source: Cloke and Park (1982).

Table 7.3: Concentration of recreation

A summary of the findings of recreation surveys	
By area	Within 50 miles of home a few locations take most of the brunt.
By car	Vast majority arrive by car and do not stray more than 100 yards from it.
By time	Notably between 3.30 p.m. and 5.30 p.m.
By season	Notably June, July, August and September.
By motive	Vast majority enjoy pleasure driving most and a picnic second.
By day	Major peak on Sunday, minor peak on Saturday.

Source: Gilg (1979).

more profitable use of the farms' resources. Indeed, in France this has been widely recognized in the very successful 'Gites' scheme.

However, the main impact of tourism has been felt in the European Alps where tourism has led to a dramatic reversal in population decline, major changes in the socioeconomic structure of local communities, and often very intensive physical development in the landscape (Mallet, 1978). In a rare survey of local residents' attitudes, the Kariels (1982) have found that in spite of the many negative aspects ofen ascribed to tourism, the overall evaluation of local inhabitants was strongly positive, although at the same time they wanted the landscape to retain its traditional character and felt that family life had been weakened by tourism. On the whole though, mass tourism on the alpine scale must inevitably destroy much of the existing culture, including local languages (White, 1974) but the greatest impact is probably felt in areas where another type of tourism has developed, namely second homes.

Second homes

Although there is no generally agreed definition of a second home (virtually every study provides its own) it is methodologically very difficult to locate them, and the numbers of homes fluctuate wildly from year to year (Shucksmith, 1983). It is generally agreed by the plethora of studies made of the phenomenon that: (a) they have been rapidly growing; (b) they are most common in Europe (see Table 7.4); (c) they are strongly related to the upper income bracket (Bielckus *et al*, 1972); (d) they are concentrated in the more scenically attractive areas, and within these a contagious process can lead to further concentration (Thissen, 1978); (e) they tend to form a further ripple of urbanization beyond the immediate commuter belt of big cities (Boyer, 1980); and (f) they are also relatively and increasingly absolutely important in upland and mountainous areas (Cribier, 1973).

Less is known however about the impact of second homes on the local environment, and although there has been a good deal of speculation about their impacts, as shown in Table 7.5, there are few hard or agreed empirical data to back this up. For example, De Vane (1975) concluded that their economic regional multiplier was roughly the same as hotels, but that insufficient data were available to assess their effect on house prices. In contrast, Ashby *et al*. (1975) asserted that one of the principal effects of the increased demand for second homes was to inflate prices. It is even harder to assess the social impact, and although Bollom (1978) has argued that local attitudes to second-home owners depend on the structure of the local community, and that antipathy to

Table 7.4: Second homes in Europe and North America

Country	% of households	Country	% of households	Country	% of households
Sweden	22	Austria	8	USA	4
Norway	17	Switzerland	8	Canada	4
Spain	17	Belgium	7	West Germany	3
France	16	Finland	7	Netherlands	3
Portugal	10	Luxemburg	6	United Kingdom	3
Denmark	10	Italy	5	Ireland	2

Sources: DART (1977) and Cribier (1973).

Table 7.5: Costs and benefits of second homes

Heading	Benefits	Costs
Acquisition	(a) Injection of capital investment (b) Employment of solicitors, estate agents	(a) Effect on local house prices (b) Effect on rented accommodation
Improvements	(a) Employment in building industry (b) Indirect employment creation (c) Housing stock improved (d) Increases rate revenues (e) Renovation of derelict buildings can improve landscape	(a) Depletes stock of cheaper housing (b) Diverts builders from new building
Expenditure	(a) Injection of consumer expenditure (b) Indirect employment creation	(a) Creates low paid seasonal jobs (b) If dwellings were permanently occupied consumer expenditure would be greater
Services	(a) Little use of private services but rates paid often for public services not consumed e.g. education	(a) Services withdrawn through underuse (b) More public services needed for peak season
Social consequences	(a) May revive village life	(a) Disintegration of communities (b) Threat to rural way of life (c) Polarization of local residents and incomes
Recreation and conservation	(a) Provides relaxation and recreation for the minority	(a) Visual degradation may result from poorly built or located second homes

Sources: Shucksmith (1983) and Pacione (1984).

second-home owners ironically tends to decline with increasing percentages of second homes, another study (Downing and Dower, 1972, 32) has argued that:

> While second homes consist merely of properties no longer wanted as first homes, they cause relatively few problems; but when they grow beyond this, politial, social, economic and environmental problems arise.

In conclusion, it could be that the problems have been exaggerated, and indeed one study of second homes in the southwest of England, the second most important area after Wales, concluded (South West Economic Planning Council, 1975, vii) that: 'second homes do not pose a problem that is widespread in the region nor are they likely to do so'. This happy state of affairs could well continue if recent trends towards the construction of purpose-built houses (as the stock of existing property is bought up), and more especially towards time-share developments, which spread the economic benefits over a wider season and ownership, continue into the late 1980s and early 1990s (White, 1978). Nonetheless, there is as yet not enough evidence to decide whether second homes are a 'curse or blessing' (Coppock, 1977). In the meantime one way in which the impact of second homes and other recreational uses could be more effectively assessed could be the development of theoretical models and statistical relationships, and so attention is now turned to this topic.

Theoretical and quantitative approaches to rural recreation

One of the most simple concepts in rural recreation studies has also been one of the most useful and enduring, namely the 'leisure budget'. This divides the use of time into essential activities (sleeping, eating, etc); work; travel to work; and the residue, leisure time (Martin and Mason, 1976). Within this residue, which now totals about 40 hours out of the 168 hours in a week, the detailed choice of leisure depends on the availability of three further resources, money, space and facilities, and also of course personal preference as shown in Figure 7.1.

The most easily quantifiable, if however the most elusive, value in these factors is money, and so not surprisingly a good deal of attention has been paid to this variable. For example, there has been a lot of evidence that petrol prices have a marked effect on rates of rural recreation (Shucksmith, 1980b), since the demand relationship is not only stable, but is also highly price-elastic. To achieve these conclusions, Shucksmith used perhaps the most widely used method of 'demand curve analysis' based on monetary values, the so-called 'Clawson method' (Clawson, 1959). The basis of the method is a three-step process of ascertaining the relationships between: (a) visitor rates and distance; (b) visitor rates and cost; and (c) number of visits to the site and an assumed set of admission charges. Although the Clawson method has been widely used and adapted (Everett, 1979), and other monetary or cost-benefit approaches have been tried, for example, regression analysis and consumer interviews of willingness to pay (Curry, 1980), the method still has a number of underlying weaknesses. For example, it assumes an entrance fee for resources that are by and large free, and it ignores not only the intervening opportunity of other sites but also personal preference, which is likely to be a key variable in a basically non-economically motivated activity.

Personal preference has been examined in a number of statistical ways. For example, Everett (1978) used multiple regression analysis to analyse 2,816 questionnaires concerning people's wildlife preferences, and found that these were positively related to the number of colours and brightness of different species. In another multivariate approach Roome (1983) used principal components analysis to conclude that professional visitors to nature reserves were positively attracted by fragile species, but amateur visitors by the access provided, and in an interesting attempt to reconcile the different expectations of different types of visitor, Price (1979) used questionnaires to derive satisfaction curves representing a range of tastes. Perhaps the most innovative approach however has been Brotherton's (1982a) attempt to relate the frequency of visits with the rate of site return as shown in Figure 7.3b. This not only provides an excellent summary of recreational use (since it is based on a 1977 home-interview survey of 5,040 respondents) (Countryside Commission, 1982) but in addition it also provides an explanation for any deviation from the curve, as shown in Figure 7.3a theoretically, and in Figure 7.3b in practice. These deviations also allow any imbalance between supply and demand, a central concept of much recreational work, to be assessed.

One aspect of this relationship is mobility and Cracknell (1967) has examined the potential supply of access provided by minor roads around one major city, Leicester. However, the advent of motorways has extended the range of the day trip, and Edwards and Dennis (1976) using gravity models and regression analysis have predicted that motorways could increase the number of longer-distance day trips to the southwest of England by between 11 and 17 per cent. In another use of the gravity model, Colenutt (1969) tried to widen its power by incorporating behavioural variables. Nonetheless, reviews of the whole gamut of recreational traffic models in rural areas, including trip generation models, distribution models, direct demand modelling, traffic assignment models, system simulation (Miles and Smith, 1977), and gravity models (Baxter and Ewing, 1981) have pointed out the following deficiencies which still need to be remedied by further research on:

1   Differing levels of road accessibility;
2   Effect of fuel prices and other costs;
3   The relative attractiveness of different sites;
4   The function of the journey itself;
5   Multiple purposes of any trip (not just to visit one site);
6   The size and type of zones to be used and the time periods to be employed; and
7   The different types of demand, effective, deferred, and potential (Countryside Commission, 1970).

Another variable to be considered, as already pointed out, would be social characteristics, and Settle (1980) has used both multiple regression analysis and the so-called 'linear logistic model' to produce a close fit between sex, car licences and age, and the number of anglers. This is not the only model to produce a close fit and in conclusion it seems that Lavery's (1975, 198) assessment of the use of models in forecasting recreation demand is still very relevant for the mid to late 1980s:

> Although our understanding of the whole and individual elements of the recreation system is very imperfect, we do know the main activities which generate the greatest growth and demand for space and we can identify current and potential conflicts with other land use activities.

Nonetheless, these broad relationships remain permissive rather than causal and, although rural recreation studies have produced remarkably concurrent findings, as outlined earlier in this chapter, while these relationships remain permissive it is still fair to ask if rural recreation geography has come of age.

Rural recreation geography: An emerging or mature discipline?

Owens (1984, 174) has commented that the 1970s saw a hiatus in recreation research as a period of evaluation replaced the previous period which had been characterized by:

> The accumulation of an ever-increasing number of empirical case studies which

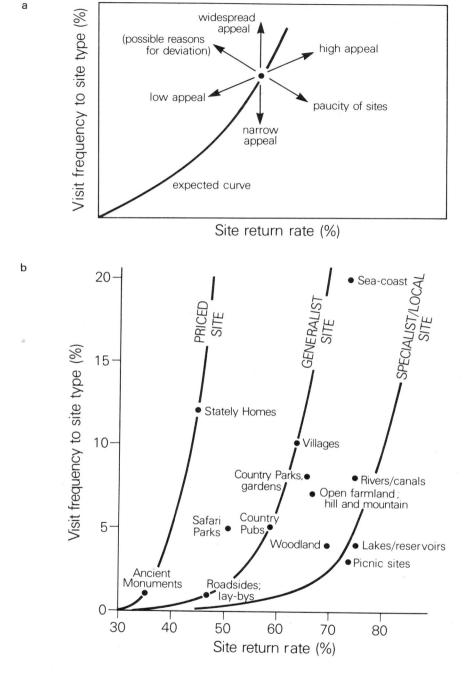

Fig 7.3 Theoretical (upper diagram) (a) and actual (lower diagram) (b) relationship between site return rate and visitor frequency. Source: Brotherton (1982a)

together comprise(d) a broadening data base to describe users, activities and patterns of use in relation to the existing supply of opportunities.

This conclusion can be confirmed by a study of the books published in the early 1970s. For example, Coppock and Duffield (1975) employed the 'leisure/resources' approach outlined above by Owens in a British context. Simmons (1975) in a world-wide review also employed this approach, and in another text, Lavery (1974) argued that there was no uniquely geographical approach. It must now be asked if the same could said in the mid-1980s.

In helping to answer this question two direct comparisons are available. First, Patmore (1983) has broadly replicated the 'leisure/resources' approach of his pioneering (1970) text, and the multi-author text *Land and Leisure*, Fisher, Lewis and Priddle (1974), and Doren, Priddle and Lewis (1979) has also used this approach in the American context. There therefore doesn't seem to have been much of a change, and furthermore Smith (1983) in his text specifically entitled *Recreation Geography* has also followed this model, albeit in the modified form of the relationship between 'travel/resources' which he calls the two main branches of the tree of recreation geography. However, Smith also points out that there is another division in the subject, between the logical positivist (objective) and the phenomenological (subjective) approaches, and that this split is delaying the development of a rural recreation geography paradigm.

In conclusion, Owens (1984) has commented that the 1980s have been characterized so far not by significant conceptual developments, but by a further widening of the empirical base. The fact that no powerfully theoretical recreation geography has emerged, however, should not be seen as a defeat for the subject, for the empirical work that has been produced has not only been very useful in its own right, but has also shown a significant mismatch between leisure aspirations and recreational resources. Attempts to reconcile these imbalances, as well as to deal with other rural problems, have formed the basis for the work of a growing number of rural geographers in the field of rural planning. Accordingly, these rural planning issues are the theme of the last two main chapters of this book

8

Land use and landscape

For many decades studies of rural land use and landscape were at the heart of all rural geographical work, and land use and landscape studies are still of intrinsic interest to many rural geographers. However, in recent years the methodological changes outlined in Chapter 1 have not only reduced the relative amount of work in this field, but also transferred the focus of interest away from the fairly limited concept of land use towards not only the more abstract concept of landscape, but increasingly to behavioural studies of how we value land and landscapes. Accordingly this chapter is divided into *land use studies, landscape studies*, and *land and landscape evaluation*.

Land use studies

In spite of the fact that there has been a very long tradition of work on rural land use, including the invaluable land use survey of the 1930s (Stamp, 1962), and its revival in the second survey of the 1960s and 1970s* (Coleman *et al*, 1974) and much other detailed work from other sources (Hart, 1980; Coppock, 1960a; Best, 1981) as shown in Table 8.1, there is still a desperate need (Hall, 1974, 414) for a 'national Domesday book for land use, preferably updated every 10 years at the time of the population census'. This statement still remains true today despite the valiant and continuing voluntary efforts of the Land Decade Educational Council to produce a national land use survey by the late 1980s (Balchin, 1981).

So the first problem facing any study of land use is the paucity of good and compatible data, for as the upper section of Table 8.2 shows, there can be considerable variations between different estimates, although the data do provide broad answers to the main points of interest defined by Peters (1970) namely: How much land is there? how is it distributed between the major land uses? and how have the allocations changed over time (and space)?

Nonetheless, Peters's second and third questions also highlight the three main problems encountered with all land use work: first, the multiplicity of data sources shown in Table 8.1; second, the different temporal and spatial units employed by these sources; and third, and perhaps most important of all, the different classifications involved. In spite of the fact that a national land use classification has been produced (DOE, 1975) it is very unlikely that the future will be any better than the present, and that national land use statistics will continue to be (Dickinson and Shaw, 1978, 299)

---

* Both are discussed more fully in Chapter 2.

Table 8.1: Land use within Britain: A summary of the principal data sources

Source and author	Area of cover	Date of data	Classification	Basis
1st LUS[+] Stamp	England and Wales	1931–34	Formal	Field survey
2nd LUS[+] Coleman	Britain	1960s and some 1970s resurvey	Extension of 1st LUS	Field survey
Best	Britain	1947–77	Functional	Development plans and MAFF[++] Census
Fordham	United Kingdom	pre-1939 to 1974	Formal	Sampling from OS[+++] Maps
DOE[++++]	Urban areas of England and Wales	1968–69	Formal	Air photos
Anderson	England and Wales	1960s	Formal	Sampling from MAFF[++] land classification maps

Sources: Rhind and Hudson (1980) and Anderson (1977).
[+]   Land Utilization Survey.
[++]   Ministry of Agriculture, Fisheries and Food.
[+++] Ordnance Survey.
[++++] Department of The Environment.

a meaningless amalgam of figures based on different classifications applied to dissimilar areal units with varying degrees of precision.

One of the most serious differences in the classifications that have been employed so far have arisen from whether to divide land up by its *use* (formal) or by its *activity* (functional). Although both the National Land Use Classification (DOE, 1975) and Dickinson and Shaw (1978) have argued for an *activity* (functional) based classification, most geographical studies have employed *use* (formal) classifications, as shown in Table 8.1, since these are easier to ascertain in field surveys or from remote sensing sources. However, as Table 8.1 also shows, these are not the only sources available. For example, Coppock (1978) has produced a comprehensive list of sources, and Rhind and Hudson (1980) have combined these into four main sources:

1   data from lists and texts;
2   data from maps;
3   data from remote sensing; and
4   data from ground surveys.

These are now considered in turn.

As Table 8.1 shows, Best (1981) has used *lists and texts*, in particular the agricultural census, to demonstrate that Britain is not excessively urbanized as shown in Table 8.2, and to argue that more space should be allocated to homes and trees. Anderson (1977) has used systematic point sampling from the Ministry of Agriculture's Land Classification *maps* (see the last section of the chapter for a commentary) to produce land use estimates very similar to Best's as shown in Table 8.2, and in the uplands, Parry *et al.* (1982) have used old Ordnance Survey maps to show that there has been extensive reclamation of

Table 8.2: Land use structures of England and Wales, Europe and USA

	Agriculture	Woodland	Urban	Other
England and Wales %				
Anderson's estimate (1960's)	79.1	6.9	9.7	4.3
Best's estimate (1961)	79.1	6.9	9.9	4.1
Best's estimate (1971)	76.9	7.5	11.0	4.6
Coleman's estimate (1963)	80.1	8.0	10.8	1.1
National Parks	87.7	8.4	1.3	2.6
Areas of Outstanding Natural Beauty	77.1	11.3	3.6	8.0
Highland England and Wales	82.5	7.1	7.3	3.1
Lowland England and Wales	77.7	6.4	10.8	5.1
Europe and USA %				
France	61.2	28.3	4.9	5.6
West Germany	54.2	29.5	11.8	4.5
Ireland	70.0	4.0	1.5	24.5
Italy	66.2	21.0	4.2	8.6
Netherlands	64.6	8.8	15.0	11.6
USA	56.7	23.2	2.7	17.4
UK	78.2	7.9	8.0	5.9
Europe Ha/1,000 people				
France	623.5	289.1	50.3	56.6
West Germany	222.3	121.2	48.3	18.5
Ireland	1619.5	91.6	34.5	566.1
Italy	360.4	114.4	22.6	47.0
Netherlands	165.2	22.5	38.3	29.6
UK	338.6	34.4	34.5	25.7

Sources: Anderson (1977), Parry (1983), Anderson (1980), Best (1979) and Jackson (1980).

moorland on the plateau uplands of the North York Moors, Exmoor, Dartmoor and the Brecon Beacons. *Remote sensing* for agricultural land use has already been discussed in Chapter 2, but for wider land use purposes, the Department of Environment (DOE, 1978b) has used air photos to define accurately the extent of all built-up areas in England and Wales. But the future almost certainly lies with multi-spectral images from satellites. In this field the LANDSAT series of satellites provides an enormous amount of data (Harris, 1979). Although a good deal of success has been achieved with regard to mapping vegetation (Hathout, 1980), there are still problems with resolution and cloud cover (Allan, 1980) and although these are gradually being resolved by radar, the problem of more detailed and fine-grained interpretation will remain for the foreseeable future (Deane, 1980). Although high technology might have appeared to make *ground surveys* redundant, they still have a very useful role, particularly when it is the function of the land use rather than its appearance that is important (Coleman *et al*, 1974), and when the process of detailed land use change is being analysed (Blacksell, 1981).

Indeed ground-based land use surveys have still provided almost all the usable knowledge we have about our land use base. For example, the irregular sample surveys conducted by the Forestry Commission (1983b) provide a mass of information on woodland age structure, species composition and ownership as well as the increase in the forested area from 6.2 to 7.3 per cent in England, from 6.8 to 11.6 per cent in Wales, and from 7.2 to 12.6 per cent in Scotland, between 1947 and 1980. However, this source suffers from the long time gap between censuses, and so more detailed and more frequent information on changing land use within the forest still has to rely on *ad hoc* surveys. For instance, Peterken and Harding (1975) from air photos and ground surveys have shown how modern forestry has reduced the numbers of 'poor colonizer' species. Harkness (1983) from land use maps, Ordnance Survey maps and air photos has shown how increased forestry has affected the landscape of National Parks. At the same time Brotherton (1983) using documentary evidence has demonstrated that the rate of afforestation has been three times greater outside the National Parks.

Within the National Parks, all land use surveys have shown how their landscapes have been much altered in recent years by the advance of agriculture and forestry (Leonard, 1980) although as Table 8.2 shows, both National Parks and Areas of Outstanding Natural Beauty (AONBs) still contain very few urban areas (Anderson, 1980). Interestingly, AONBs have more woodland but less farmland than National Parks.

Attempts to provide land use data for other countries are bedevilled by even worse problems of definition and classification than those already outlined, but Best (1979) has tried to harmonize European national statistics to produce the figures shown in Table 8.2, and in the USA although Jackson (1980, 6) has argued that: 'it is almost impossible to provide adequate statistics on land use . . . because of differing definitions', he has nonetheless provided the estimates shown in Table 8.2 using Department of Agriculture statistics. However, the difficulties are highlighted by the high percentage of other land shown in the table, 17.4 per cent, which is largely accounted for by swamp, desert and bare rock. Similar problems are found in Canada, but in both countries there is a growing realization that the land resource base is not limitless and that more information is needed on land use change (McCuaig and Manning, 1982).

In spite of the difficulties already discussed and the limited information available, a number of general conclusions can be made about land use and land use change. First, it is essential that the two-way flow of land use change should be emphasized, for as both Figures 8.1 and 8.2 show, net gains can conceal a far more complex picture. In addition, the rate of change can be exaggerated, because it is often the same pieces of marginal land that are being continually transferred from one use to another, and then back again. This is emphasized by Coleman (1969), who has advocated a departure from the traditional classifications of agricultural land, forestry and so on, and instead produced a concentric ring model radiating out from townscape to urban fringe, to farmscape, to marginal fringe, and finally to wildscape.

Second, it can be concluded for both the UK (Best, 1976) and for the EEC (Best, 1979) that although there are major differences within countries, notably in the relative areas devoted to woodland and urban land, that there is a common pattern of land loss to both urban use and woodland as shown in Figure 8.1. However, because of the recent advent of food surpluses, this isn't the same problem as it would have been a few years ago. In the USA, a similar

Fig 8.1   Land use changes in Britain, Europe and North America. The top diagram shows land use change between 1933 and 1980 for the UK, the middle diagram shows flows between land uses between 1960 and 1980 for the UK, and the bottom diagram shows land use changes in Europe and North America between 1961 and 1971. Sources: Parry (1983) top two diagrams and Champion (1983) bottom diagram.

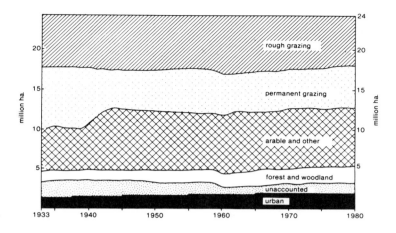

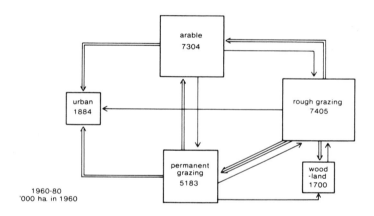

1960-80
'000 ha. in 1960

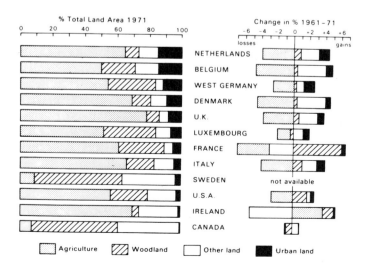

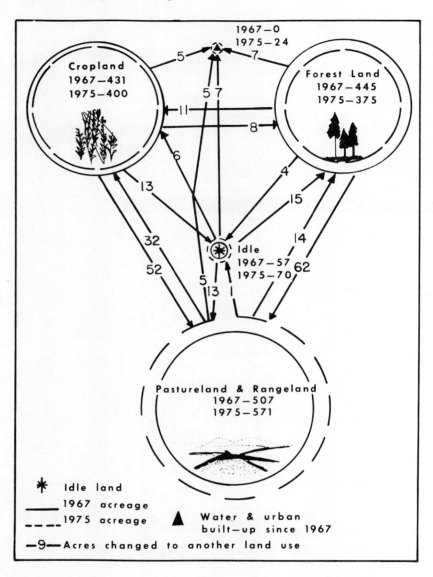

Fig 8.2   Land use change in the USA. Figures shown are millions of acres. Source: Jackson (1981)

pattern of land loss to urban use is also observed (Jackson, 1980), but a major difference is a loss of forest land and farmland in equal amounts to other extensive uses, notably recreation and conservation.

Although studies of land use like those outlined above provide useful overall data, they only produce a limited picture of what the countryside actually looks like and how it is changing in appearance, and this type of study remains at the heart, if not the mind, of many of even the most socioeconomic quantitative geographers.

Landscape studies

The study of landscape is one of the oldest sub-disciplines within geography, but during the quantitative revolution of the 1960s and early 1970s such work was roundly criticized for its descriptive nature and its search for uniqueness. For a while, landscape work was restricted to historical geography, where it produced, and continues to produce, a rich vein of national (Hoskins, 1955) and regional (Armstrong, 1975) descriptions of the evolution of landscape. This book is, however, confined to contemporary change, and by the mid-1970s concern about landscape change and particularly the loss of traditional features led to a revival of interest in the field, particularly by those geographers employed in the public services. Indeed the main studies in both the two major landscape types, the *lowlands* and *uplands*, have been conducted by public agencies, and these provide the essential evidence for the rest of this section.

In the *lowlands* the basic facts of land use have already been outlined in Table 8.2 (8th line of data) and although there have been a number of local studies of landscape change (Blacksell and Gilg, 1981), only nationally commissioned studies can provide a large enough resource base for an adequate survey of both landscape change and the behavioural aspects behind the change. Fortunately, the Countryside Commission has not only commissioned one such study (Westmacott and Worthington, 1974) but was wise enough to commission a re-study in the early 1980s (Westmacott and Worthington, 1984).

The initial study compared the landscapes of 1945 and 1972 and examined farmers' motives and attitudes to landscape change in seven 'type' areas: intensive arable; extensive arable (2); dairying; mixed; general cropping; and livestock rearing, dairying and mixed. As Table 8.3 and Figure 8.3 show the most dramatic changes were the enlargement of field size and the loss of hedgerow trees, with a much greater rate of change in the arable areas in order to take advantage of large modern machinery. More subtle but equally important changes were the cleaning and straightening of water courses, the ploughing up of herb-rich grassland, and the construction of modern farm buildings, out of scale and out of character with the traditional landscape. The main reasons for the changes were economic and the increasing redundancy of certain features like hedgerows.

This pattern of change wasn't, however, completely repeated in the 1970s, according to the 1980s re-study, which compared the landscapes of 1983 with those of 1972 as shown in Table 8.3. First, the re-study noted a marked slowing down in the rate of change, partly because there was little left to change, particularly in the arable areas. Second, there appeared to have been an evening-out of change, with the mixed/pastoral farming areas recording similar rates of

Table 8.3: Changes in lowland agriculture landscape 1945–72–83

		Cambridgeshire (intensive arable)	Huntingdonshire (extensive arable)	Dorset (extensive arable)	Somerset (dairying)	Herefordshire (mixed)	Yorkshire (general cropping)	Warwickshire (livestock and mixed)
Average field size (hectares)	1945	5.5	7.5	7.5	3.5	4.5	6.0	5.0
	1972	13.5	18.5	8.5	5.5	6.5	8.0	6.5
	1983	13.5	21.5	10.0	6.5	6.5	9.0	7.5
Increase in field size %	1945–72	128	137	16	44	45	26	25
	1972–83	1.5	16.5	19	14.5	0	10	22.5
Length of hedges removed (metres/hectare)	1945–72	35.5	28.0	6.0	15.0	14.5	11.5	10.0
	1972–83	0.5	9.0	5.0	4.5	0.5	2.5	10.5
Length of hedges remaining (metres/hectare)	1972	—	44.0	63.6	94.7	149.2	71.2	125.8
	1983	56.0	34.5	58.0	90.0	148.0	68.0	114.5
Number of hedge-row trees (per 40 hectares)	1947	39	59	17	50	49	51	69
	1972	5	12	7	15	40	33	81
	1983	4	6	9	15	32	27	29
1972 as % of 1947		13	20	41	30	82	65	117
1983 as % of 1972		69	53	123	99	81	83	35
Age distribution %*								
Mature	1972	11	31	39	30	34	50	22
	1983	14	14	21	30	37	54	31
Semi-mature	1972	60	31	16	60	37	34	50
	1983	11	4	16	9	16	16	32
Sapling	1972	23	18	43	9	27	5	23
	1983	74	67	60	60	40	20	30
Dead	1972	6	6	2	1	2	11	5
	1983	1	1	3	1	7	10	7

Source: Westmacott and Worthington (1984).   * Trees

Sketch showing a view similar to that in the Somerset study area in 1945

Sketch of the same view, as predicted in *NAL*

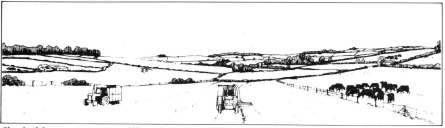

Sketch of the same view, as it would have changed since 1972

Fig 8.3 Changing landscapes in the lowlands. Hypothetical landscapes, showing a typical Somerset landscape in 1945, (top diagram) how it might have changed between 1972 and 1983 (middle diagram), and how such a landscape has actually changed between 1972 and 1983 (bottom diagram). Source: Westmacott and Worthington (1984) Reproduced with the permission of the Countryside Commission

change to the arable areas, partly because arable farming had markedly expanded into these areas during the period from 1972 to 1983. Third, major landscape changes appeared to be more and more confined to a change in either tenure or farming type, and were clearly the result of well-thought-out decisions rather than the often piecemeal changes of the 1960s. Fourth, and in some places a reversal of the previous trend, the number of hedgerow trees had either declined less rapidly or even started to increase (apart from major losses due to Dutch Elm disease as in Warwickshire), and in particular the number of saplings had markedly increased.

One major reason for this slowing down was the developing 'social conscience' of farmers, in reaction to increased criticism of their actions. But at the same time, the economic arguments in favour of larger fields have continued to grow, and Sturrock and Cathie (1980) have calculated that an increase in field size from 10 to 20 hectares will give an 11 per cent increase in work rate with a

3-metre wide implement, and a 24 per cent increase with a 12-metre wide implement.

The official reaction to these changes (more fully discussed in Chapter 9) has been to try and persuade farmers to adopt a more conservationist attitude, notably on unproductive land (Leonard and Cobham, 1977). But other commentators have called for the introduction of planning controls over farming operations and have branded the changes as the 'theft of the countryside' (Shoard, 1980).

In the *uplands*, the basic facts of land use have already been outlined in Table 8.2 (7th line of data), but these data ignore the distinction between farmland and rough pasture, which is particularly important in Scotland, where much of the so-called agricultural area is moorland. They also fail to show the steady decline in the amount of grazing land as shown in Figure 8.1 and the complex two-way flows between uses also shown in Figure 8.1. Indeed, unlike the lowlands where much of the change is due to a rejection of the 200-year-old 'enclosure land-scape' much land use change in the uplands is the result of a continuing revaluation of the economic benefits, or otherwise, of reclaiming moorland for agricultural use, and abandonment of farmland (Parry, 1976) can take place along-side reclamation (Parry *et al.* 1982). It is therefore very important to try and assess the more enduring changes shown in Figure 8.4, and Table 8.4.

Within the post-war period, the two most enduring changes appear to have been a marked transfer from rough pasture to either woodland or farmland, and Parry *et al.* (1982) have shown that about three-quarters of the change has been to woodland, as shown in Figure 8.4, and that this pattern is fairly geographically even, except for Dartmoor as shown in Table 8.4. Other studies have shown that much more land could be converted to forestry, to give a possible 30 per cent cover to produce (ITE, 1978, 31): 'a dramatic landscape change' and a landscape: 'similar to that of Sweden'.

However, this isn't likely to occur within the National Parks and although, as Table 8.2 shows, they have a higher percentage of woodland than the rest of the uplands, voluntary agreements to limit any further afforestation will reduce their relative share. Table 8.4 has already shown that most National Parks lost more rough pasture to woodland than to agriculture, but in the southwest Table 8.4 also shows that Dartmoor lost more rough pasture to farmland than to woodland, and in a study of land use change in the nearby Exmoor National Park, Lord Porchester (1977) found that the moorland area had been reduced, from 24,000 hectares in 1947 to 19,000 hectares in 1976, with 3,800 hectares

Table 8.4: Enduring conversion of rough pasture

Study area	Rough pasture area 1980	To farmland	To woodland	To other uses	Total (hectares)	Total area
Mid-Wales	72,400	9,700	23,200	610	33,500	—
North York Moors[+]	50,900	5,500	12,500	40	18,000	141,568
Northumberland[+]	71,500	2,400	12,500	—	14,900	103,082
Brecon Beacons[+]	65,800	3,800	6,400	110	10,300	134,500
Snowdonia[+]	63,000	1,000	3,900	80	5,000	212,000
Dartmoor[+]	49,300	1,400	700	160	2,300	94,535

[+] National Park areas.
Sources: Parry *et al.* (1982) and Woods (1984).

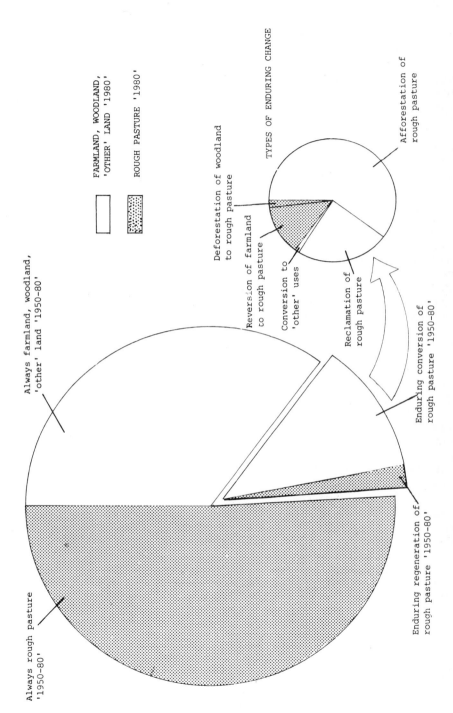

FARMLAND, WOODLAND, 'OTHER' LAND '1980'

ROUGH PASTURE '1980'

Always farmland, woodland, 'other' land '1950-80'

Always rough pasture '1950-80'

Deforestation of woodland to rough pasture

Reversion of farmland to rough pasture

Conversion to 'other' uses

Reclamation of rough pasture

TYPES OF ENDURING CHANGE

Afforestation of rough pasture

Enduring conversion of rough pasture '1950-80'

Enduring regeneration of rough pasture '1950-80'

Fig 8.4 Land use change in the uplands. Source: Woods (1984) Reproduced with the permission of the Countryside Commission

going to farmland but only 1,200 hectares going to woodland.

Although the studies outlined above provide useful information, they fail to produce a detailed picture of landscape change or the reasons behind these changes. Fortunately, in parallel with the re-study of lowland landscapes, the Countryside Commission has also commissioned a similar study of upland landscapes in both a full (Sinclair, 1983) and abbreviated form (Allaby, 1983). Both volumes adopt the lowlands study formula and examine the changes and the reasons behind them in 12 'type areas' ranging from northeast to southwest England as shown in Table 8.5. In addition, more attention is paid to the socio-economic structure of farming and in particular the greater reliance of hill farming on government and EEC grant aid and subsidies.

These grants have encouraged changes in hill farming which have led to a loss of semi-natural vegetation, traditional farm woodland and field boundaries, and have been offset by increases in commercial forestry, and cultivated crops and grass. The main reasons for these changes are said to be economic pressures, grant aid, and a gradual replacement of older, smaller, and more conservative farmers with younger, larger and more profit-orientated farmers. Indeed, only 29 per cent of farmers had actually altered field boundaries between 1949 and 1978, and only one-third had reclaimed moorland in the last 7 to 15 years, and so the picture that emerges is one of a more dynamic profit-motivated minority replacing a more conservation and landscape-orientated majority where still 56 per cent of farmers are over 50.

In addition to the major landscape changes already outlined, Sinclair (1983) has also found a small increase in the amount of enclosed land (1.8 per cent), but a major change from vernacular enclosure to fences, with the area enclosed by walls and hedges down by 13 and 37 per cent respectively between 1949 and 1978, but the area enclosed by fences up by 45 per cent, to give the overall pattern shown in Table 8.5. Another change has been extensive draining of moorland (Stewart and Lance, 1983) and this allied to increased sheep numbers and other factors (Anderson and Yalden, 1981) has led not only to a decrease in moorland vegetation but to a reduction in wildlife, notably grouse.

Wildlife changes are the final element in the landscape equation, and the changes outlined above (in both uplands and lowlands, as a result of the increased use of machinery and also chemicals) have led to the destruction of 95 per cent of herb-rich meadow, 80 per cent of lowland sheep walks, 40 per cent of lowland heaths, 30–50 per cent of ancient lowland woods and 50 per cent of lowland fens and mires since 1945 (NCC, 1984), and therefore the extinction or rapid decline of much flora and fauna.

Two key features in the landscape change process, as the above discussion has shown, are farmers continuing revaluation or reclassification of land, and the behavioural attitude of farmers to change. However, farmers aren't the only group to value land and landscape and so the last section of this chapter examines the varying ways in which land and landscape can be assessed by other groups.

Land and landscape evaluation

Traditionally land has been valued mainly for its intrinsic worth as a factor of production, but with the growth of recreation and tourism, land in the form of

Table 8.5: Changes in upland landscapes 1967–78

	% Net change in rough pasture	% Net change in farmland	% Net change in woodland	Characteristic form of vernacular enclosure	% of enclosed area in vernacular form
Alwinton (Cheviots)	− 12	+ 26	+ 113	Walls	4
Lunedale (North Pennines)	− 3	+ 35	+ 66	Walls	75
Shap (Lake District)	− 5	+ 4	+ 41	Walls	63
Bransdale (North York Moors)	− 5	− 5	+ 100	—	—
Heptonstall (Northern Peak)	+ 0.3	− 0.2	—	Walls	90
Moynash (Southern Peak)	− 21	+ 0.8	+ 50	Walls	100
Llanfachreth (Snowdonia)	− 3	+ 8	+ 1	Walls	85
Ysbyty Ystwyth (Mid-Wales)	− 1	+ 7	—	Hedges	5
Glascwm (Mid-Wales)	− 20	+ 14	+ 38	—	—
Glyntawe (Brecon Beacons)	− 13	− 0.7	+ 229	Hedges	1
Lynton (Exmoor)	− 13	+ 8	− 5	Hedgebanks	82
Widecombe (Dartmoor)	− 9	+ 17	− 18	Hedgebanks	98
All areas	− 7	+ 9	+ 36	—	67

Source: Woods (1984).

landscape can be just as, if not more, valuable a resource. Both land and landscape are however very difficult to value.

In the case of land, its market price is of course one yardstick, but this can be distorted by all types of factors, for example, proximity to expanding towns. Therefore, virtually all methods of land evaluation have limited themselves to the physical factors of climate, relief and soil. An example of this approach is provided by the Ministry of Agriculture's 'Agricultural Land Classification' maps which use rainfall, transpiration, temperature and exposure as the key climatic indices; altitude, slope, and surface irregularities as the key relief indices; and wetness, depth, texture, structure, stoniness and available water capacity as the key soil indices (Agricultural Land Service, 1966). These indices have been used to produce 113 maps for England and Wales at a scale of 1:63,360 and although the five grades shown on the maps provide a broad pattern of where the best (grade 1) and worst (grade 5) land is, the maps have, however, been roundly criticized. For example, the width of the five grades is not consistent, and grade 3 not only accounts for about half the land, but also covers a much wider range of possibilities than, say, grade 1 (Gilg, 1975a), and even the subsequent sub-division of grade 3 in the 1970s cannot be applied objectively (Worthington, 1982). Oher criticisms relate to the size of units used (a minimum of 80 hectares) and the fact that the system is based on the flexibility of land and not on its productivity or capability (Boddington, 1978). These types of criticisms can, however, be extended to most methods of land classification and evaluation, and so Flaherty and Smit (1982) have proposed a broader-based system in which other parameters including housing, recreation and conservation are included.

Similar types of comments can also be made concerning landscape evaluation, and once again the methods employed are very dependent on the purpose of the evaluation. For example, Figure 8.5 shows that this can vary from landscape improvement to recreation, and that the two main spatial units have been kilometre squares and landscape tracts (Penning-Rowsell, 1975). The most common practical uses of landscape evaluation have however been in the preparation of land use plans (Penning-Rowsell, 1974) planning in rural areas (Rogers, 1981b) and the definition of areas of scenic value like Areas of Outstanding Natural Beauty (Anderson, 1981b; Preece, 1980).

In spite of the fact that landscape evaluation techniques only date from the mid-1960s a voluminous literature has evolved. Penning-Rowsell (1981) has provided an excellent review of the literature during this period, and has concluded that it saw a sustained attempt to devise techniques for the quantification of the scenic quality or value of landscapes. Penning-Rowsell has also concluded that these were inspired by a growing awareness of landscape deterioration, through either urban expansion or inadequate management, and a perceived need for increased protection of those areas of greatest landscape quality. However, most of these methods paid insufficient attention to the highly complex nature of landscape values, while overemphasizing scenic attractiveness. Furthermore, excessive emphasis was placed both on measurable landscape attributes and on devising universal methods, while neglecting the social attitudes, feelings and values of the general public.

In more detail, Penning-Rowsell (1981) has defined three clear stages in the development of landscape evaluation: first, an early emphasis on *intuitive and morphological* approaches; second, the development of complex *statistical* approaches; and third, an emphasis on public *preference*.

The first, *intuitive and morphological* approaches (Linton, 1968) were heavily biased in favour of land form, and were divided into two stages. An initial objective inventory of landform and land use, and then a qualitative assessment of the value of each type of landscape so identified. However, these approaches tended to use geomorphological concepts of landform, and only recently has research examined how people actually see landforms, by using artists' sketches (Killeen and Buyhoff, 1983) and computer-generated models to produce artifical landscapes (Guldmann, 1980). Hull and Buyhoff (1983), however, used the more traditional medium of colour slides to show that the further the horizon the greater the perceived scenic beauty.

In a search for a greater relationship between landscape and numbers the early to mid-1970s produced a set of more complex *statistical* approaches. These not only involved far more variables, but also secondary data sources in addition to field survey. In the first and most quoted example of the genre, the Conventry–Solihull–Warwickshire study (1971) set the mould by using multiple regression analysis as shown in Figure 8.6. This study attracted a good deal of interest and replication (Blacksell and Gilg, 1975) and in a massive appraisal of the approach (Robinson *et al.*, 1976) the method was modified to standardize the preference scores in step B of Figure 8.6, and a factor analysis was made of the measured landscape variables in Step A, so that criticisms of the statistical

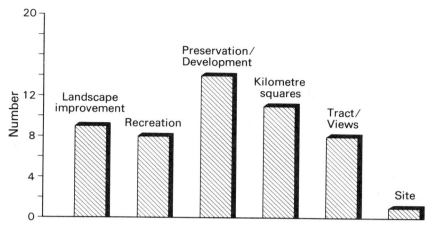

Fig 8.5 Different purposes and units of data collection for landscape evaluation. Source: Gilg (1979)

validity of the method could be overcome. Nonetheless, this latter modification has been rejected (Dearden and Rosenblood, 1980) and Dearden (1980) has tried to develop the multiple regression model by making the dependent variable more representative, and the independent variables more comprehensive. In another multivariate approach, Shuttleworth (1980a) has used principal components analysis to show that relief accounts for 20.5 per cent of the variation in landscape description produced by a nine-point semantic differential scale.

However, few of the statistical methods have achieved more than a 70 per cent explanation rate, and furthermore, the regressions have often been based on very small samples of professional people, and the mechanistic techniques have cloaked what is essentially unquantifiable with an indefensible professional mysticism. In a reaction, therefore, to these problems the last few years have seen two major changes in approach. First, a return to viewing the landscape in totality, rather than the sum of its component parts, and second, a much greater emphasis on the *preferences* of the general public, rather than professional people, even though some studies have shown quite a good correlation between the two (Preece, 1980).

Nonetheless, the public *preference* approach is not without its problems, notably how to assess people's preferences, and whether a consensus could or should be achieved. Two problems arise with assessment. The first is how to represent the landscape, since assessment in the field is too laborious and costly. Although slides and photos have been the traditional medium, and are widely accepted as substitutes for the landscape (Shuttleworth, 1980b), Kreimer (1977) has also argued that they need to be put more fully into their context. The second is how to elicit people's preferences, and though score sheets, questionnaires and semantic difference scales have been widely used, the search is still on for a more effective method (Propst and Buyhoff, 1980).

With regard to the search for consensus the results are often contradictory. For example, Penning-Rowsell (1982), in a survey of 540 people, found a general consensus which was related to familiarity with the area and also age, but

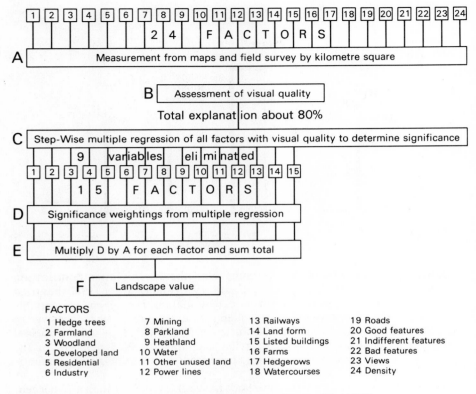

Fig 8.6   A multivariate method of landscape evaluation. Source: Gilg (1979)

unrelated to social class and sex. However, Wellman and Buyhoff (1980) have found no regional familiarity effect, Lyons (1983) has found an age differential, and Buyhoff *et al.* (1983) have found that although different nationalities broadly agree over landscape quality, select national groups produced a greater degree of similarity.

Nonetheless, a substantial degree of agreement has been achieved, not only between different demographic groups, but also between the whole range of approaches, *intuitive, statistical* and personal *preference* (Briggs and France, 1980) and it can generally be concluded that good landscapes will have substantial relative relief, a major water body, diverse land use, and some historical artifacts, but will lack modern industry and major transport facilities.

This still leaves the fundamental question of how people value landscapes, and how different tastes and needs can be catered for. Accordingly, there has been a gradual realization that the endeavour should advance over a broader field rather than trying to seek a 'universal' method (Dearden, 1981), that subjective methods are probably more cost-effective (Price, 1976), that any method should include room for different tastes (Jacques, 1980), and that much could be learnt from the approach of experts in the area of aesthetic and artistic judgement (Carlson, 1977).

Indeed the relationship between landscape and the arts has been extensively explored in recent years, and Olwig (1984) Pocock (1981) and Relph (1981) have shown how many of our values are derived from such sources, notably humanism and literature. Perhaps the most stimulating approach however has been Appleton's (1975) brilliant development of prospect–refuge theory. This theory, which has been developed from an extensive survey of landscape paintings, argues that landscapes are highly symbolic habitats in which the most valued landscapes allow us to see (prospect) without being seen (refuge) and therefore relate to our origins as hunter–gatherers.

However, landscapes can now be seen in a variety of ways, in pictures, from the air, and in maps (Johnson and Pitzl, 1981) and this adds further to the list of problems yet to be solved in Lowenthal's (1978) excellent essay on finding valued landscapes which include varieties of taste, the effect of social milieu, the distinction between landscape and place, differences between public and professional preferences, whether tastes are innate or learnt, the effect of literary and historical factors, season, time of day, viewpoint chosen and direction of view, novelty and familiarity, distance and memory lending enchantment, personal sensitivity and intensity of feelings, and the effect of experience and training. It is not surprising therefore that Lowenthal (1978, 389) has concluded that:

> landscape evaluation studies neither ascertain landscape tastes nor assess public preferences; they tell us nothing about what landscapes the public values or why.

This is maybe one reason why increasing attention is being turned to methods by which the general public can (a) 'read the landscape' (Muir, 1981), (b) have it interpreted for them (Countryside Commission and CCS, 1975); or (c) even interpret their own everyday landscape (Meinig, 1979) rather than having experts trying to value the invaluable.

9

Rural planning and land management

Previous chapters have examined rural affairs from a broadly policy-free viewpoint, but within each chapter it has been fairly apparent that many of the topics have been affected by planning measure, for example, long-term forestry policy in Chapter 3, key settlement policy in Chapter 4 and rural transport policy in Chapter 6. It is not the purpose of this book to examine the whole gamut of rural planning policies, since these have been comprehensively examined elsewhere (Davidson and Wibberley, 1977; Gilg, 1979; Burnell-Held and Visser, 1984) but instead, to provide a commentary about those policy areas where rural geographers have been most active, not only as researchers, but also as advocates of policy changes – either by sitting on government committees or very commonly by being employed by government agencies.

Nonetheless it remains very difficult to isolate the precise impacts of planning procedures on rural areas (Cloke and Hanrahan, 1984), partly because rural planning measures are very diffuse and emanate from no single agency or piece of legislation, and also because in many cases they are mutually exclusive or even contradictory. In spite of this it is possible to outline a number of basic themes in rural planning. For example, Davidson and Wibberley (1977) have divided rural planning into activities (planning for the development of agriculture, forestry and leisure) and resources (planning for the protection of resources, wildlife and landscape). Other aims for rural planning over the past 40 years are shown in Table 9.1 and from this it can be argued that three main roles for rural planning can be defined: a *resource development role*; a *conflict resolution role*; and a *land management, recreation and conservation role*. The relationship of these roles is now considered in outline, before discussing them in more detail in the rest of this chapter.

Rural planning roles

Although the overall aim of rural planning (Wibberley, 1984, 1) should be 'the creation of conditions under which rural activities can flourish and resources be sustained', sectional interests have ensured that some activities have flourished more than others, and that the fortunes of each of the three roles outlined above has fluctuated over the last 40 years or so.

However, for most of this period the *resource development* role has been prominent, especially in the context of agricultural policy, where the UK, and the USA have instituted massive farm-support measures to boost farm incomes and or production levels. In the last decade however, the almost blind adherence to agricultural expansion policies has been increasingly questioned, as the side effects of agricultural development and a decline in other less subsidized areas

Table 9.1: Aims for rural planning 1944 to 1977

Orwin's 4 aims (UK)
1 To develop employment opportunities in the countryside;
2 to improve living conditions;
3 to ameliorate the handicap of small population size for socio-economic life; and
4 to preserve the amenities of the countryside and the beauty of the rural scene.

Thorn's 4 aims (UK)
1 To reconcile the competition for land between agriculture, housing development, industry and recreation;
2 to preserve the beauty and character of the countryside;
3 to redistribute the population into viable villages both economically and socially; and
4 to give people a greater say in the future of the countryside.

Lassey's 8 aims (USA)
1 To preserve ecological integrity in order to provide a continuing supply of life-supporting resources;
2 to ensure efficient and appropriate land uses;
3 to provide healthy living conditions;
4 to provide an aesthetically pleasing environment;
5 to provide effective social, economic and governmental institutions;
6 to improve human welfare;
7 to ensure that new buildings and adapted landscapes are of a pleasing design; and
8 to be comprehensive.

Sources: Orwin (1944), Thorns (1970) and Lassey (1977).

of the rural economy have become apparent (MacLeary, 1981). Furthermore, it has become increasingly obvious that the support policies are going to those who need them least, and in the USA it has been estimated that almost half the $2 billion spent on deficiency payments in 1978 went to the 10 per cent of farmers who were the largest producers (Dinse, 1983).

Although agricultural support policies still take the lion's share of resource development budgets, programmes for employment and social development have become an increasingly important part of the rural resource development equation, particularly in the remoter and more upland areas, where although farming is heavily subsidized, it cannot provide a sound rural economy on its own.

The wider rural economy is one of the main concerns of the second main role of rural planning, the *conflict resolution* role. For many years the core of the problem was that most of the countryside was, and still is, controlled by one group, the farmers, who have acted in a way that has often conflicted with the interests of other members of society, the vast majority of the population (Wibberley, 1972). Not surprisingly, these people have increasingly formed themselves into pressure groups either to fight individual issues (Leat, 1981) or increasingly to play an important role in policy formulation (Lowe and Goyder, 1983). But in recent years the conflict has widened out, and is no longer simply one between farming and urban-based interests, but of one power group against another, and of one value system against another (Cherry, 1976).

Furthermore, governments have aided this process by sectionalizing ministries and agencies so that their policies are geared towards the achievement of individual goals in agriculture, forestry, recreation and nature conservation (CRC, 1976). One solution that has often been proposed to this sectional interest problem is the creation of a Ministry of Rural Affairs (Wibberley, 1976), but both the Conservative government (Gilg, 1984b) and the Labour opposition (Labour Party, 1981) have rejected this idea in the 1980s, although the Liberal Party (1983) has proposed a parliamentary committee to review the impact of government policies on the countryside.

One problem that any such rural minister would face is that there is no consensus about the kind of rural environment that we want rural planners to provide (Doyle and Tranter, 1978), although there have been repeated calls for a committee to be set up to find out if such a consensus can be achieved (Gilg, 1978).

On one central issue, however, a clear consensus seems to have emerged, and that is that a sectional approach to rural planning is no longer acceptable, and produces more conflict than it resolves. Since a merger of interests at the governmental level seems unlikely, there has been a growing interest in achieving integration at the grass-roots level by promoting *land management* policies. The initial impetus for these came from the need to reconcile recreation and conservation in protected landscapes, especially in National Parks, and this approach has since been expanded to embrace the concept of management agreements in agriculture and forestry, and has found its fullest expression in the growing concepts of multiple land use and integrated rural development, much favoured by the European Economic Community.

The rest of this chapter explores the three roles outlined above in more detail and in particular looks at the nature of the problem, the policy response to it, and the effectiveness of this response particularly from the point of view of rural geography.

Resource development roles

Agricultural policy
The background to the agricultural policy problem is extremely complex and diverse with a number of interconnected variables. It is generally agreed however that agricultural policies are needed, and indeed adopted by virtually every developed country in the world (OECD, 1977) because of the following factors. First of all, there is a need to overcome fluctuations in the weather and wild swings between gluts and shortages, and to promote long-term confidence. Second, there is a need to exploit comparative advantage, for example, milk in mild wet areas, and conversely also a need to subsidize production in otherwise marginal areas. Third, there is a need to boost agricultural production and thus the national economy, since agriculture, although it employs fewer and fewer people directly, accounts for a relatively high percentage of GNP, and employs at least 10 per cent of the workforce indirectly, and furthermore, food prices are still the single most important item in most people's budgets. Fourth, and finally, there is a need to promote national security, keep the balance of payments under control (Cmnd 6020, 1975) and yet at the same time encourage world trade in both food and non-food items. As Figure 9.1 shows, however, these aren't the only goals of agricultural policy, but they are the most common, as the following review of agricultural policy responses in the UK, Europe and the USA will show.

Within the UK there have been three main periods of policy (Tracy, 1976). The first was a period when *laissez-faire* methods were first abandoned, initially during the agricultural recession of the 1930s, when only partial policies were

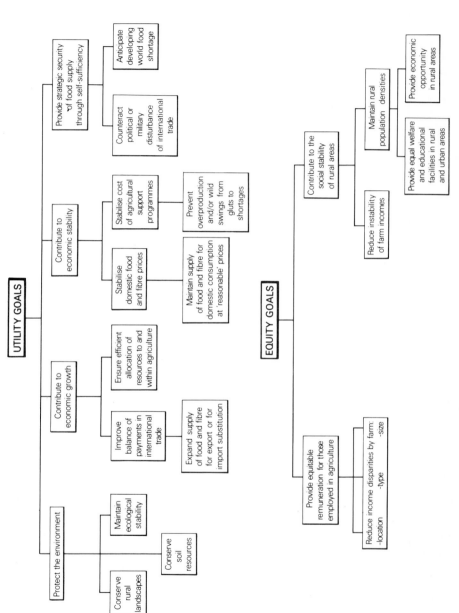

Fig 9.1 Goals of agricultural policy. Source: Bowler (1979)

introduced, only for full *dirigiste* control to be introduced in the 1940s as a response to the food shortages of the Second World War and its aftermath. The second was a period of more relaxed but expansionist policies in the 1950s and 1960s based on the 1947 Agriculture Act. And then, the third period saw its gradual replacement in the 1970s, as EEC farm policies were gradually introduced. Therefore the 1946 to mid-1970s period was dominated by the 1947 Act policies which had (Gilg 1979, 27) the aims of:

> promoting and maintaining by the provision of guaranteed prices and assured markets . . . a stable and efficient agriculture capable of producing such part of the nation's food and other agricultural produce as in the national interest it is desirable to produce in the United Kingdom and of producing it at minimum prices consistent with proper remuneration and living conditions for farmers and workers in agriculture and an adequate return on capital invested in the industry.

These aims were achieved by an annual price review which set guaranteed prices, the introduction of marketing boards and quotas, and by cash grants and subsidies to either improve the agricultural resource base or to subsidize farming in marginal areas. It is generally agreed that these measures greatly aided the impressive improvements made in both agricultural production and productivity throughout the post-war years (Winegarten and Acland-Hood, 1978), and that the fundamental aims of the policy were to increase production (Bowman *et al.*, 1976), and that this should continue to be the case even after accession to the EEC (CRC, 1978; Cmnd 7458, 1979).

Within the EEC, Article 39 of The Treaty of Rome gives the Common Agricultural Policy (CAP) the following objectives:

(a) to increase agricultural productivity by promoting technical progress and by ensuring the rational development of agricultural production and the optimum utilization of the factors of production, in particular labour;

(b) thus to ensure a fair standard of living for the agricultural community, in particular by increasing the individual earnings of persons engaged in agriculture;

(c) to stabilize markets;

(d) to assure the availability of supplies, and

(e) to ensure that supplies reach consumers at reasonable prices.

The policy uses similar methods to those employed by the 1947 Act in Britain, namely a two-pronged approach, with first, policies to support the price of farm produce via market regulation, and second, policies to improve the structure of farming, but as Table 9.2 shows, market regulation takes by far the biggest share of the budget. There is, however, a major difference in the method of price support, since the CAP uses a combination of intervention buying, tarriff barriers, and export subsidies to keep prices high for its farmers. The combination of high support prices (until 1984 without any quotas on production) and a protected market has certainly increased production in the EEC and led to three major problems within both the CAP and the EEC. First, massive surpluses have built up, as shown in Table 9.3, second, the CAP takes up well over half the EEC budget, and within the CAP, the price support policy takes up well over three-quarters of the CAP budget leaving little left for structural reform (Commission of the EC, 1977), and third, no common European policy has yet emerged.

The problem of surpluses isn't however unique to the EEC (European Communities, 1984) and apart from a brief period in the mid-1970s, agricultural

Table 9.2: Pattern of public expenditure on agriculture

£ million	1981/82
Price guarantees and production grants	23.6
Support for capital and other improvements	167.4
Support for agriculture in special areas	102.8
Market regulation under CAP*	678.6

* Broad pattern of market regulation by the Intervention Board for Agriculture.

1981	UK expenditure in £ million borne by Exchequer[+]	1982
72.6	Cereals purchases	129.9
22.8	Skimmed milk	70.9
156.0	Total**	298.0
(127.0)	Less receipts from sales	(76.4)
29.0	Net expenditure	221.6

1981	UK expenditure in £ million borne by CAP[+]	1982
283.7	Export refunds (outside EEC)	255.3
86.3	Sheep variable premium	112.0
24.5	Whisky refund	81.0
79.6	Export refunds (inside EEC)	77.5
(124.4)	Import levies	(126.5)
630.2	Total**	737.9

( ) indicates receipts.
[+] Only expenditure above £70 million in 1982 shown.
** Total includes smaller items not shown.
Sources: Cmnd 8804 (1983) and Cmnd 8963 (1983).

Table 9.3: EEC intervention stocks and consumption

Commodity (tonnes)	Total	stocks	Total stocks as a % of consumption
Wheat	EEC	1,157,000	3.2
	UK	310	negligible
Butter	EEC	447,469	26.6
	UK	59,744	14.8
Skim milk	EEC	455,439	227.7
powder	UK	21,289	26.6
Olive oil	EEC	81,000*	15.9
Table wine	EEC	9,000,000*	9.8

* Hectolitres.
Source: HL Select Committee on the European Communities (1980).

policy in the USA has had to deal with surpluses rather than shortages. Therefore the goals of US agricultural policy have been rather different and have been justified (Pacione, 1984, 76) in terms of:

1 improving the farming community's share in national economic prosperity;
2 the belief that a plentiful supply of food should be provided to everyone at reasonable prices;
3 the importance of agriculture to the balance of payments; and
4 the general view that agriculture and the family farm are components of American life worth preserving.

It has also been argued that under such policies that not only do producers benefit from price and income support, but that consumers also benefit from ample supplies of food at reasonable prices (OECD, 1983). In common with UK and CAP policies, US farm policies have been divided into (a) structural measures to improve basic farm resources, and (b) a variety of farm price and income support measures. These have ranged from measures to dispose of surpluses, methods to raise the price of commodities in the market place, and acreage allotments or production quotas under the 1933 Agricultural Adjustment Act. This Act remained the main piece of legislation, till food shortages in the mid-1970s forced a change in direction, away from controlling, towards increasing production. However, by the early 1980s surpluses had emerged again and measures to control production, including acreage limitation, and a payment of between 80 and 90 per cent of any production foregone, replaced measures to boost production, in the Agriculture and Food Act of 1977 (OECD, 1983).

One major reason for the reappearance of US surpluses was the growth of surpluses of temperate foods within the EEC and attempts to sell them on the world market. However, as Tables 9.4 and 9.5 show, the EEC has been far more effective in putting its house in order, since it has not only shed far more farm labour, but spends less per head on farm workers, and yet still remains the biggest single food importer (mainly of tropical foods and also because the UK remains a major food importer from the old Commonwealth), while the USA remains the biggest single exporter.

Increasingly therefore, agricultural policy is no longer nationally, but internationally based, and a number of rural geographers have recognized this trend. For example, Tarrant (1980a) has shown how the failure to produce an international European currency has considerably distorted the CAP, and Tarrant (1980b) has also shown how the mismatch between supply and demand in the milk sector has continued to cost the community dear. However, the most comprehensive summary of the effects of the agricultural policy measures shown in Table 9.6 has been produced by Bowler (1979) who concluded that:

1   agricultural policies have had adverse effects on world trade, especially for developing countries;
2   the policies have also induced unwarranted increases in production;
3   domestic product prices have been distorted in relation to import prices;
4   significant welfare and budgetary costs have resulted from protectionist policies; and

Table 9.4: Key agricultural data for EEC and USA

	EEC	USA
Farmland in use (million hectares)	93.4	428.8
Farm work force in millions		
1968 (EEC) 1970 (USA)	12.07	3.46
In 1979	7.89	3.30
Difference	4.18	0.16
Amount spent in income support per person employed in agriculture, 1979	$1,441	$1,760

Source: Anon (1982).

Table 9.5: World trade in farm products

1979 %	Imports	Exports
EEC	25.7	9.8
Japan	13.4	0.7
USA	11.8	18.6
Canada	2.3	4.7
Australia	0.6	4.1
New Zealand	0.2	1.6
Rest of world	46.0	60.5

Source: Anon (1982).

5 the income objective is being reached for only certain sections of the farm population.

To these conclusions could be added the severe impact of agricultural intensification on both landscape (already discussed in Chapter 8) and conservation (to be discussed later in this chapter) and also the effect on rural communities. For example, in the USA, Buttel (1983) has concluded that improvements in farm productivity and farm structure have set in motion a downward multiplier, or spiral of decline, in rural-community economic activity, and in the UK, Slee (1981) has argued that agricultural policy has had the same effect, especially in the marginal areas. In particular, agricultural policy is sectoral, while other policies, especially regional employment policies are spatial. The net result is that one policy, farm policy, provides grant aid to shed farm labour while another policy regional aid, attempts to redeploy this labour. Slee therefore argues for a more comprehensive and less contradictory policy, aimed at a more balanced level of resource development. Accordingly, the next section examines some of the attempts to provide plans for rural development outside agriculture.

Planning for employment in rural areas
The nature of the rural employment problem has already been fully examined in Chapter 5 although, as Figure 5.7 shows, many lowland and accessible areas are now doing relatively well in employment terms. However, many of the more remote and upland areas are still experiencing population and employment decline, especially away from key settlement villages and growth-pole towns, and there are a number of reasons, including regional balance and existing social capital (Gilg, 1976) why government policies should try and retain population and employment in these areas.

At the outset it should be made clear that rural development policies are not synonymous with agricultural policy (Commins and Drudy, 1980), and indeed it can be argued that undue emphasis has been placed on agricultural development in rural areas, to produce strong agricultures but weak rural economies (Wibberley, 1981). In a similar vein, it has also been argued that too much attention has been paid to manufacturing industry, as a panacea to the problems of rural regions (Drudy and Drudy, 1977). Nonetheless, as the following review of rural development agencies demonstrates, manufacturing industry has been the centrepiece of much of their work.

In the UK, regional policies based on grant aid, loans and other mainly fiscal measures have attempted to divert industry away from the more prosperous towards the less fortunate areas since the 1930s. These policies are however not

Table 9.6: Types of agricultural policy measures

Measures designed to influence:

The growth and development of agriculture	The terms and levels of compensation of production	The economic structure of agriculture	The environmental quality of rural areas

: subsidies on the cost of inputs (fertilizers, machinery, buildings) drainage, ploughing	: demand management
	  — advertising
	  — produce grading schemes
	  — multiple price
		: financial assistance for land consolidation schemes	: pollution control measures
			: land zoning ordinances

: preferential taxes	: supply management
	  — land retirement
	  — import quotas
	  — tariff protection and import levies
		: farm amalgamation grants	

: preferential credit interest rates	— marketing quotas
	  — production quotas
	  — buffer stock schemes
	  — target/threshold prices with intervention (support) buying
		: land reform schemes	

: financial assistance to co-operatives or syndicates for the joint purchase of inputs	— export subsidies — bilateral/multilateral commodity agreements	: retirement and discontinuation pensions/grants	

: financial assistance to agricultural educational extension and research establishments	: direct payments
	  — deficiency payments
	  — fixed rate subsidies/ production grants
	  — direct income support grants
		: retraining schemes
		: financial assistance to co-operatives for production or marketing	

		: inheritance laws (land taxation)	
		: farm wage determination	

Source: Bowler (1979).

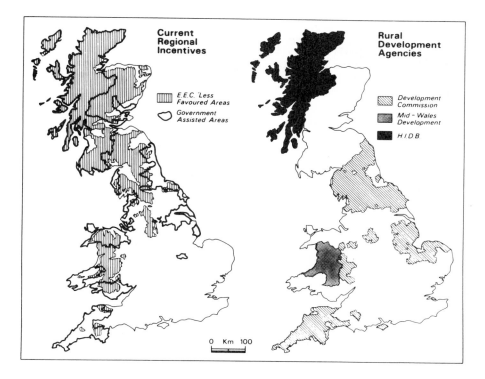

Fig 9.2 Regional incentives and rural development agencies 1983. Source: G. Williams (1984b)

specifically rural, although as Figure 9.2 shows, the Government Assisted Areas, albeit much reduced since 1979 (Gilg, 1980a) include significant rural areas in southwest England and Scotland.

The oldest specifically rural development policy dates from 1909 when the Development Commission was first set up. Whilst it was originally concerned with the development of craft industries, its remit and budget have widened considerably since the mid-1960s. In 1968 it set up an agent COSIRA (The Council for Small Industries in Rural Areas) to aid small rural industry, and then in 1975 its work was expanded to include the creation of 1,500 jobs in the Special Investment Areas, shown in Figure 9.2. These were reduced in extent and renamed Rural Development Areas in the early 1980s (Gilg, 1985), by the Development Commission (1982) which has defined them as the areas where:

1 unemployment is above the average for Great Britain;
2 there is an inadequate or unsatisfactory range of employment opportunities;
3 population decline or sparsity of population has an adverse effect;
4 there is a net outward migration of people of working age;
5 the age structure is heavily biased towards elderly people; and
6 there is poor access to services and facilities.

The net result of all these changes is that the Development Commission's work is now divided into four main areas. First, devising and implementing rolling programmes of small-scale factory/workshop building in rural towns of less than 15,000 population. Second, providing through COSIRA a business management, technical advice and credit service to small manufacturing and service industry employers of less than 20 skilled men, in towns generally under 10,000 people. Third, the encouragement of voluntary initiatives and pioneering rural community experiments through its active support of 36 county-based Rural Community Councils (RCCs), and fourth, requesting local authorities, both County and District Councils, to prepare 'Action Plans' denoting those settlements possessing good prospects of recovery (G. Williams, 1984b). Furthermore, most rural councils have their own policies for rural development (ADC, 1983).

Since 1975 however the Development Commission's spatial remit has been reduced to England only. In Scotland, the Highlands and Islands Development Board (HIDB) was set up in 1965 (over the area shown in Figure 9.2) to promote not only regional social and economic development and a more rational land use strategy, but also to integrate the region with national economic growth. In its early days, it concentrated on a growth-pole strategy focused around large-scale, highly capitalized projects, for example, pulp mills and aluminium smelters, but with the failure of both these projects in the early 1980s (Armstrong, 1982) it has become increasingly concerned with both the remoter areas within the region and with the stimulus of indigenous potential.

In Wales, three separate bodies have been created over the years to deal with the special problems of central Wales. In 1957 the Mid Wales Development Association was set up to promote inward investment, and with the aid of Development Area status in 1966, and Development Commission advance factories, it followed a growth-pole approach based on Newtown and outside business interests. In 1977 the Association was replaced by the Development Board for Rural Wales and then in the early 1980s this was replaced by Mid-Wales Development (for the area shown in Figure 9.2), and although both bodies continued to invest over half their resources in Newtown, they both began to spread investment more widely over the region after the mid-1970s (DBRW, 1982)

Both the HIDB, with a staff of 250 and a budget of £30 million, and Mid-Wales Development, with a staff of 90 and a budget of £10 million, have powers of land-acquisition/disposal, the ability to erect buildings, and the power to make grants and loans at preferential rates to potential growth industries and commercial ventures. But the absence of any real power over agriculture and forestry in mid-Wales is, however, a serious limitation on the development of comprehensive land use and economic policies there (Minay, 1977), in contrast to HIDB which, as Table 9.7 shows, spent about 10 per cent of its budget between 1973 and 1982 on land development in agriculture and forestry. However, both agencies promote a small but significant programme of non-economic grants and loans to strengthen social and economic infrastructure. In conclusion, the central focus of British agency activity is undoubtedly employment creation by the use of special areas, the provision of premises and industrial sites, grant aid, loans, advice and training, help with housing and community aid.

In contrast, the EEC has no specific rural policy, although there are three programmes for creating rural employment (HL Select Committee, 1979). The first of these is the CAP, and although its main thrust is agricultural price support, a 'less favoured areas directive' allows special aid to be given to remote

Table 9.7: Expenditure by development agencies

A	*Development Commission* %	74/75	78/79	81/82
	COSIRA	70	44	33
	Advance factories	9	44	46
	Community development	14	7	12
	Miscellaneous	3	2	5
	Administration	3	3	5
	Total (£ million)	4.4	13.1	12.1
	Factory/workshop approvals (No.)	7	279	155

B	*Highlands and Islands Development Board*	(£ m.) Grants	(£ m.) Loans/Shares	Employment Created	Retained
	Sector (1973–82)				
	Land development	6.8	14.1	1,077	700
	Fisheries	9.0	28.5	1,597	578
	Manufacturing/Processing	14.4	21.1	6,289	2,301
	Construction	2.1	4.1	1,258	499
	Tourism	30.4	13.9	3,711	923
	Other service industries	6.9	7.8	2,430	697
	TOTAL	69.6	89.5	16,362	5,698

C	*Mid-Wales Development* %	78/79	80/81	81/82
	Administration	8.5	16.5	15.6
	Housing construction	34.0	19.5	17.2
	Industry and Commerce	46.0	53.0	49.4
	Marketing	4.0	5.8	8.7
	Social development	6.5	5.0	5.2
	Research	1.0	0.3	N/A

Source: G. Williams (1984b).

and mountainous areas (the British areas are shown in Figure 9.2), and socio-economic advisers can give advice and aid to the growing number of farmers seeking to supplement their farm income by tourism and other activities. The second programme is represented by both the Regional and Social Funds, but they have been of very limited significance for the remoter rural areas (HL Select Committee, 1979). The third programme is as yet exploratory and involves the establishment of three integrated rural development programmes in the Western Isles of Scotland, the department of Lozere in France and the province of Luxembourg in Belgium. This programme involves the integration of agriculture with other projects and although as yet in its infancy the early signs are encouraging (HIDB, 1983).

Interest in the concept of integrated rural development has also been expressed in the USA (Fuguitt, 1979) particularly with regard to preventing further population decline in the more remote and upland rural areas. Contemporary measures to reduce sharp geographical disparities in economic opportunity and living standards in the USA date from the 1960s, when an area economic development programme was introduced with the possibility of setting up cross-border multi-state planning agencies (Estall, 1982). During the 1960s, five such agencies, known as 'Regional Planning Commissions', were established in the South West (Arizona, Colorado, Utah and New Mexico), in the upper Great Lakes, the Ozarks, the Coastal Plains of Carolina and Georgia, and New England, with a special region being set up for Appalachia. The regional plans for these areas, however, overemphasized highway building, and non-road spending had to be spatially concentrated, thus creating difficulties. Furthermore, the sums of money involved were quite small, and measured against the scale of the needs of the regions and other budgets, the $260 million spent over the first 10 years of the operation was (Estall, 1982, 36): 'a drop in the ocean'. However during the 1970s, pressure to achieve balanced natural growth led to proposals for a nationwide system of Regional Commissions; but with the change in administration in 1980 to an anti-welfare, anti-planning government, these have now been abandoned. In addition, some commentators saw the commissions as 'excrescences' on the existing system of federal and state agencies, and probably most important of all, they were seen to have largely failed in their task of promoting the economic development of lagging regions in a cooperative multi-state planning context.

In Britain however, regional planning policies have been more politically acceptable, at least till the Thatcher years, and they have been seen to achieve a greater degree of success. For example, Table 9.7b shows that the HIDB claims to have created over 16,000 jobs between 1973 and 1982, and although Bryden (1981) has estimated that this is an exaggeration by at least a factor of 2 (since many of the jobs might have been created without aid) he also concluded that HIDB policies have had a significant and positive impact on the region. Nonetheless, Armstrong (1982) has argued that the landed estates, which dominate the area, have been reluctant to invest in the region even with HIDB aid, and accordingly argues that the policy will continue to fail until there are radical reforms in land policy and much more comprehensive powers are made available to develop the entire primary resource base.

In Mid-Wales it has been estimated that employment increased by 12 per cent and the population decreased by 0.7 per cent between 1961 and 1975, but that without the development policy, that employment would have decreased by 2 per cent and population by 7 per cent (Broady, 1980). Nonetheless, the policy has been criticized for favouring too few growth-pole areas, notably Newtown, and it has been concluded that the policy has not yet succeeded in generating the demographic structure which would be required for self-generating growth.

In the wider context of Wales however, Cooke (1980) has concluded that Welsh Development Agency funds have been spent as wisely as possible, although the tendency to shy away from large enterprises, after a number of failures has perhaps led to too great an emphasis on 'seed-bed' firms which will take a long time to solve the problem. Finally, for England, G. Williams (1984a) has argued that the advance factory programme has been too haphazard, in spite of the new criteria for Rural Development Areas already outlined.

In conclusion, it would appear that rural employment policy has been too circumscribed by growth-centre options, an over-emphasis on manufacturing, and a failure to take an integrated enough approach. However, all these views are changing in the 1980s and there is a growing awareness of the need to (a) include service employment in regional policy (Cmnd 9111, 1983), (b) to examine the effect of new technology (NCVO, 1980b) and (c) to embrace integrated rural development (HIDB, 1983).

Conflict resolution roles

Land use planning and urban development in the countryside

Most developed nations have created systems to control land use in the post-war period for a variety of reasons. In the UK the traditional reasons for land use planning have been to protect farmland from urban sprawl, to provide a balance between regions, to protect landscapes, and to keep conflicting land uses apart. Over the years the relative weight of these arguments has fluctuated, and much emphasis is now placed on reconciling conflict, and the government (DOE, 1980, 1) has stated that:

> The planning system balances the protection of the natural and built environment with the pressures of economic and social change. The need for the planning system is unquestioned and its workings have brought great and lasting benefits.

Unfortunately, no method of balancing competing claims for land use has yet been found, although a number of methods including cost-benefit analysis (Wibberley, 1959) and goals-achievement matrices have been tried.

In the USA land use planning has been slow to evolve and can still be said to be in its infancy (Burnell-Held and Visser, 1984) because of a strong ideological bias against government infringement of public property rights, and because the huge area of the country, combined with its relatively sparse population, has much reduced conflict (Lassey, 1977). Indeed the main area of conflict in American land use history has been (Clawson, 1975b, 473) 'the interplay between the private forces pushing for land development and the public forces pushing towards a degree of social control over land use'. Because the USA is a federal country with a multitude of agencies and levels of government, this has led to uncoordinated, fragmented and inconsistent land use policies (de Neufville, 1981). Nonetheless, four basic criteria are usually employed by counties when planning for rural areas: halting urban sprawl; minimizing expensive public services; preventing damage to environmentally sensitive areas; and the preservation of open space, or in the words of Lefaver (1978, 8): 'The main purpose of planning is to exclude urban-type development from rural land'.

This is a theme echoed elsewhere in the developed world, and the OECD has concluded that although farm production losses due to urban expansion may appear negligible, and may easily be compensated by technical advances, that the hugh still unsatisfied world need for food still means that the use of fertile land for non-agricultural purposes must be regarded as intolerably wasteful (OECD, 1976).

Indeed, the need to protect farmland for a food-rationed Britain was one of the main thoughts uppermost in the government's mind when it passed the first major British policy response to land use conflict, the 1947 Town and Country Planning Act. This Act was modified in 1968, when rigid Development Plans based on detailed land use plans were replaced by more flexible Structure Plans based on diagrams. The 1947 and 1968 Acts were consolidated in 1971 to produce the statute under which modern land use planning derives its power, the 1971 Town and Country Planning Act (Cullingworth, 1982).

Under this Act, planning is basically divided into two functions, plan making and development control. Under the present system of plan making, County Councils prepare Structure Plans which are supposed to show (Housing and

Local Government, 1970) the broad patterns of settlement change, industrial and employment development, communications, and landscape conservation over the next 10 to 20 years, as shown in Figure 9.3. The plans have however attracted a good deal of criticism for being too physical and not giving enough attention to socioeconomic affairs, for being too broad, and yet also for being too detailed, for being contradictory in their policies, and for taking too long to be prepared (Barrass, 1979) in spite of continued government advice to the contrary (DOE, 1979). In sum, structure planning has become an incremental process arbitrating between powerful interests over specific issues and has lost any pretensions to being a long-term or comprehensive decision taking process (Cross and Bristow, 1983).

A more detailed expression is given to these Structure Plans by Local Plans (Healey, 1983). Unfortunately, these are prepared by different organizations, namely the District Councils, and they do not always follow the Structure Plan for the wider area as closely as they could. A major problem with both types of plan is that it has taken far longer than expected to prepare them, and although the Structure Plans were by and large completed by the 1980s, and were beginning to be reviewed by then (Bracken, 1980), Local Plans, particularly in rural areas, may not all be completed till the twenty-first century (Bruton, 1983).

This means that those charged with making decisions about planning applications, to develop or change land use, have had to, and will continue to, operate in many areas in a policy vacuum. A further problem is that development control is done by the District Councils and so is divorced from those planners charged with the production and review of Structure Plans. Nonetheless, development control by District Councils is the main vehicle by which land use conflicts are resolved, although it can only do this when applications are made to develop or change the use of land.

When an application for planning permission is made to a District Council the decision is made in the light of policies contained in plans for the area, the views of interested parties, and the merits of each individual case. About 80 per cent of applications are granted with or without conditions, although this high success rate may reflect the fact that many potential applicants with little chance of success are deterred from applying. An applicant may appeal to the Department of the Environment over a refusal or if he feels that the conditions attached are too onerous. In this case, a public inquiry is held to help the Department come to a decision. In addition to appeals by individuals, public inquiries are frequently held into issues of national or regional importance, and those that go beyond the normal limits or competence of local planning authorities, like the development of coal fields, motorways, airports and nuclear power. In these cases the procedure has some of the characteristics of an Environmental Impact Assessment.

In Europe, although a mass of different methods of planning have been employed, a number of common themes emerge. For example, all planning systems are concerned with the allocation of physical land use and most employ zoning methods and, following encouragement from the EEC, many countries now use Environmental Impact Assessment procedures (R. Williams, 1984). However, there is a good deal of difference in the spatial levels employed, and although many European countries follow a top-down centralized form of planning, for example Holland (Burnell-Held and Visser, 1984), others, notably Denmark and Switzerland, stick rigidly to bottom-up planning.

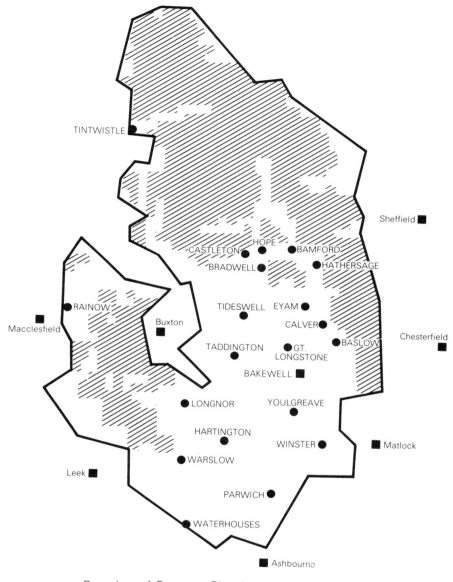

- ——— Boundary of Structure Plan Area
- ■ Major service centre
- ● Other service centre
- /// Natural zone

Fig 9.3 An example of a Structure Plan. The plan for the Peak District National Park has two main features. First, a division into a natural (upland) area and an economically active (lowland) area. Second, within the economically active area a further division into major and other service centres in line with central place theory. Source: Peak Park Joint Planning Board (1976)

This type of approach has been the one followed in the USA, and for many decades land use controls were exercised only at the local (county) level, and only in the 1970s did a number of states, particularly those with rapid population growth (Dillard, 1983), introduce state land use planning.

However, federal land use planning still remains a 'dead issue' (Healy and Rosenberg, 1980). Even at the state level, planning is mainly limited to a few areas within states and to specific issues, for instance, Hawaii is one of the few states to have a state-wide system of control, and in other states, the focus is on individual issues, for example in Florida on the protection of critical areas; in Vermont on controlling leisure-home development; and in California on protecting the coast for posterity (Jackson, 1980). At the municipality level, 'rural environmental planning' places special emphasis on aesthetics, natural area protection, wildlife habitat protection, conservation zoning, provision of recreation facilities, access to public waters, protecting agricultural land, improving water quality and controlling the rate of growth (Sargent, 1976).

This variety of purpose is matched by a similar diversity of method, and even for the purpose of retaining farmland.from urban sprawl a wide variety of techniques have been used (Lapping and Clemenson, 1984). For example, zoning of land for exclusive farm use, public purchase of development rights, public purchase of farmland, transferable development rights (rights to develop land in a no growth area sold to developers in a growth area), differential taxation of farmland, special legal and local tax freedoms in farmland areas (Berry and Plaut, 1978), deferred tax, land trusts (Lapping, 1977), and positive or negative easements (a convenant over the use of land) (Strong, 1983). Indeed such is the variety of method that Mandelker and Cunningham's (1979) text on *Planning and Control of Land Development* requires 1314 pages to deal with the great diversity. A major problem with all these methods, is that they involve large compensation payments to buy out property rights, a problem that was solved in the UK by the 1947 Town and Country Planning Act, which bought out all claims for compensation in a once and for all payment in the early 1950s.

This is one reason why many commentaries on rural land use planning in the UK have agreed that it has been generally successful in its four main areas of concern: the *urban fringe, key settlements, preservation of farmland* and *protection of landscapes*.

Within the *urban fringe*, where most land is further protected by green belt status from strong development pressures (Elson, 1981), studies in both the Birmingham (Gregory, 1970) and London green belts (Munton, 1983) have shown that the areas have been protected from urban sprawl. However, a number of institutional uses and transport facilities have been allowed, for example, the construction of the M25 Motorway around London and London airport, as well as much conversion to recreational rather than farm use.

In the area of *key settlement* policy (see Chapter 4 for background), Cloke (1979) in a study of Warwickshire (pressured rural area) and Devon (remote rural area) concluded that planning had been most effective in controlling pressure, rather than encouraging development in the less pressured areas, and that council housing had been more commonly located in key settlements than private housing. In another study of six remote or pressured areas, Martin and Voorhees (1981) confirmed Cloke's findings but argued that agricultural and amenity policies had been the main reason for the preservation of the countryside and not key settlement policies. Finally, Blacksell and Gilg (1981) in a study of Devon confirmed that development control had largely restricted new development to existing villages, but not always the chosen key settlements.

Studies of the *preservation of farmland* have also concluded that urban sprawl has been contained, but at a socal and economic price, in terms of those council tenants left in inner city tower blocks and inflated land values for private house buyers (Hall *et al.*, 1973). Nonetheless, planning has slowed the rate of land loss from around 25,000 hectares per year in the 1930s to about 15,000 hectares per year in the post war years (Best, 1977) even though the rate has been regionally uneven, with the older industrial areas taking more farmland as they rebuilt their outmoded environments to modern space standards (Best and Champion, 1970). However, planning has been less successful in diverting development away from the best agricultural land (Gilg, 1975b) and Coleman (1976; 1977) has used her second land utilization survey to argue that planning could have preserved much more farmland, and prevented much piecemeal development in the 'rurban' fringe, if more derelict land in the inner city had been used. But in recent years, as agricultural production has grown, the need to safeguard farm land has been increasingly questioned, and although Anderson (1984) concurs with Coleman that complete urban containment could have been feasible, by reducing densities from 35.6 hectares per 1,000 people to 25.6 hectares per 1,000 people, she also argues that the expansion of living standards for the mass of British people has been a preferable goal.

Another planning goal has been *landscape protection* and general surveys (Blacksell, 1979) have concluded that the landscapes of the protected areas have been preserved from many small-scale, if not large-scale intrusions when the national economic or defence interest has prevailed. More detailed studies have used nationally collected development control data, and Brotherton (1982b) has concluded that though national parks and other rural areas are subject to a greater pressure for development than urban areas, as shown in Table 9.8a that they are nonetheless being protected by a greater refusal rate. However, the use of national data of this sort is prone to a good deal of error (McNamara and Healey, 1984) and so other studies have used the raw data from planning offices to overcome in particular the problem of weighting, i.e. one application for 100 houses being treated equally with one application for one house. This (Blacksell and Gilg, 1977; Anderson, 1981a) has produced some conflicting evidence in that AONB policy appears to have had a more negative impact on new development in East Sussex than in East Devon, as shown in Table 9.8b. It also shows the major difficulty in assessing the impact of planning policies in any of the four issues outlined above, namely that planning goals are numerous, and the achievement of any one policy, for example the protection of good farmland, may conflict with the most economic location of a key settlement, and this may in turn conflict with landscape preservation. Accordingly, both conflict resolution and its assessment are very difficult goals to achieve.

Attempts to tackle this problem are in fact mainly restricted to the UK, and in North America work has been largely confined to exploratory studies of the relationships between urban expansion and population change, and economic function and soil capability (Pierce, 1981) and has so far confirmed that the relationships are more complex than previously supposed, thus replicating UK experience (Best, 1981). Other work has argued that there is no real land problem since at recent rates of urbanization, as shown in Figure 9.4, little more than 4 per cent of the USA will be urbanized by the year 2000 (Hart, 1976).

Nonetheless, even in the corn belt states, other workers have suggested it would be prudent to conserve the finite resource of farmland (Platt, 1981) even though the methods outlined earlier in the chapter, especially zoning, differential taxation and purchase of development rights, have been found to be either ineffective or too expensive in both the USA (Peterson, 1983) and Canada (Gayler, 1982) where it has been argued that in some areas of the best climate and soil in the Niagara fruit belt, little of this area could be left by the early 1990s.

In conclusion, there is little doubt that the resolution of conflicts between urban development and the countryside is a very complex issue, that there have been many different policy approaches, and that rural geographers, especially in the UK, have made both empirical and policy contributions.

Land use planning, and recreation and conservation in protected landscapes

In an increasingly urbanized world there is a diminishing supply of rural land left over for conservation and recreation purposes. Very often the same piece of land is used for both purposes, and although this land is commonly protected by some form of government agency, a clash of interests can and frequently does arise between the case for conservation and the case for recreation.

Most commentators agree that the case for conservation rests on ethics, aesthetics, cultural and scientific values, material benefits from nature, and the need to maintain an ecological balance (Green, 1981), or on the grouping of these arguments into: ethical/aesthetic; scientific/educational; and self-preser-

Table 9.8: Planning applications and protected landscapes

A Development pressures in England and Wales 1975/76

	Population millions	Number of Planning applications	Number per 1,000 people	Refusal rate %
London	7.11	47,784	6.7	13.3
Other metropolitan districts	11.67	78.571	6.7	14.2
England and Wales	49.22	484,267	9.8	16.3
Shire districts	30.18	358,854	11.7	17.3
National Parks	0.26	6,058	23.7	25.5

B Development control in Areas of Outstanding Natural Beauty

Applications	East Sussex	East Devon
Total: in study area	571	2,990
Total: in AONB	65 (11%)	810 (27%)
Residential applications only: in Study Area	392	1,305
Residential applications only: in AONB	48 (12%)	549 (42%)
Residential applications as % of all AONB applications	74%	68%
Residential applications as % of all applications outside AONB	69%	35%

Decisions	Permitted: Refused	Permitted: Refused
All applications in study area	1.3:1	2.6:1
All residential applications in Study Area	1.0:1	1.7:1
All applications in AONB	1.2:1	4.3:1
All residential applications in AONB	0.8:1	1.6:1
All applications outside AONB	1.3:1	2.2:1
All residential applications outside AONB	1.1:1	1.7:1

Sources: Brotherton (1982b) and Anderson (1981a).

vationist/disaster prevention approaches (Gilg, 1980b). In more detail the Department of the Environment (DOE, 1972) has listed 11 reasons for preserving wildlife:

1 as a contributory component of ecological stability;
2 as a monitor of environmental pollution;
3 for the maintenance of genetic variability;
4 for the provision of a source of renewable biological resources;
5 for the needs of scientific research into the environment;
6 for its cultural and recreational value;
7 as a component of the aesthetic quality for the landscape;
8 for environmental education;
9 for the economic value of its resource, scientific and recreational components;
10 to provide future generations with a wide choice of biological capital and
11 for moral and ethical reasons.

The scale of countryside recreation has already been discussed in Chapter 7 – suffice to repeat here that it is massive, and that it does conflict with conservation. This was recognized as long ago as 1974 in the 'Sandford' report on National Parks (DOE, 1974) which argued that where conservation and recreation were irreconcilable, although they often can be reconciled even in Nature Reserves (Usher and Miller, 1975), conservation should have precedence. This proposal was accepted by government in 1976 (DOE, 1976).

Attempts to plan for conservation and recreation have both used similar policies of area-based protection, and ranked hierarchies of sites, based on the related concepts of *carrying capacity* (Sinden, 1976), *potential surface analysis* (Zetter, 1974) and *island-biogeographic theory* (Margules *et al.*, 1982; McCoy, 1983). All these concepts try to relate size and habitat to the amount and diversity of either recreational use or species, that any site can support.

Accordingly, the Nature Conservancy Council (NCC, 1984) ranks sites in decreasing order of importance from: world heritage sites; to internationally important sites; to UK sites; to marginal farmland; and to poor wildlife habitats, as shown in Figure 9.5. Similarly, recreational sites can be ranked into mainly recreational areas, regional and country parks (Slee, 1982) and into multi-purpose areas like Areas of Outstanding Natural Beauty (Countryside Commission, 1983a) and National Parks, as shown in Figure 9.6. Of most interest, however, are National Parks, not only because they are the most widespread area, but also because they try to reconcile many different uses within their boundaries, and thus try to achieve multiple land use management, the theme of the rest of this chapter.

In the UK, 10 National Parks were designated in England and Wales in the 1950s with the major aims of conserving their natural beauty and providing opportunities for outdoor recreation. At first, because of their remoteness the two aims didn't really conflict, but with the increase of recreation in the 1960s, outlined in Chapter 7, it became necessary to reconcile the two uses. The main vehicle for achieving this goal was the introduction of National Park plans in the mid 1970s.

The plans are in fact 'management plans' and should not be confused with the 'land use plans' outlined in the previous section (Dennier, 1980). Some parks, like the Peak District (Peak Park Joint Planning Board, 1976), are fortunate since they prepare both types of plan, but most national park authorities have to

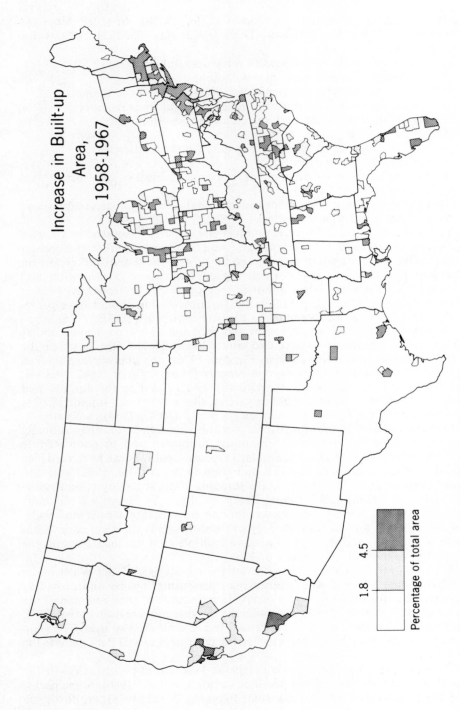

Increase in Built-up Area, 1958-1967

Percentage of total area

1.8   4.5

Fig 9.4 Increase in built-up area of the USA 1958-67. Source: Hart (1976)

take account of the land use plans for the area prepared by county and local councils, albeit often in close cooperation.

At the outset, the plans try and see conservation and recreation as complementary policies, but in detail the plans have all evolved specific policies to keep recreation away from the more sensitive areas. In general, the plans don't attempt to cater for peak demand and instead try to manage a lower demand by two forms of management; first, by access, and second, on the site. Access management involves setting up alternative attractions on the edges of the park, signposting visitors to the more robust sites, and actually closing some access points and providing public transport alternatives (Peak Park Joint Planning Board, 1978). Site management involves the gentle persuasion of visitors to use only certain parts of a site, by only providing facilities, signs and footpaths in these areas. At both the park-wide and local level, the common approach is one of dividing the area up into 'honeypots' and 'quiet areas' as shown in Figure 9.7.

However, a number of commentators have argued that national park plans have paid too much attention to recreation, and that their record with regard to conservation is negligible (Brotherton, 1982c) although it has improved since the 1974 'Sandford' report. The MacEwens (1981) however argue that attempts to conserve the national parks are still 'cosmetic' and are fundamentally hampered by a lack of money, resources and political power to do anything really worthwhile. This is disturbing since in the future the clash of interests will grow worse as recreational pursuits become more pervasive and intrusive in their impacts, for example, the development of skiing in Britain's only really wild landscape, the Cairngorms (Carter, 1981).

In the USA, although the national park authorities have much greater power in that they control the land directly and don't have to deal with the existing socioeconomic structures of a settled landscape, the 'dilemma' between preservation and recreational use is still very apparent (Pigram, 1983). In line with the UK, the 1970s saw a swing towards conservation, and a policy of recreating the environment first seen by European explorers. This has involved increasingly restricting access to the 37 national parks, an increased use of park-and-ride schemes, and zoning the parks into wilderness and recreational areas. Nonetheless, Runte (1979) has argued that too much attention has been paid to 'scenic monumentalism' and not enough to ecology, and that economic forces still prevent more land being designated even in the barren state of Alaska. Therefore the Conservation Foundation (1972, 11–13) has recommended four wider aims for the second century of national parks in the USA:

1 We recommend that the National Park system reassert its traditional role as conservator of the timeless natural assets of the United States;

2 We recommend that greatly expanded citizen participation be made fundamental to the planning and management of the Parks;

3 We recommend that the National Park system serve urban America by assuring a distinctive recreational opportunity based on natural values; and

4 We recommend that the National Park system be used as a showcase of man's proper stewardship of land, water and air.

Conservation cannot however be solely viewed in the national context, since not only do many species migrate (and recreation is increasingly an international phenomenon), but pollution, for example acid rain, crosses national borders at will. Accordingly not only must international agreements for birds (Lang,

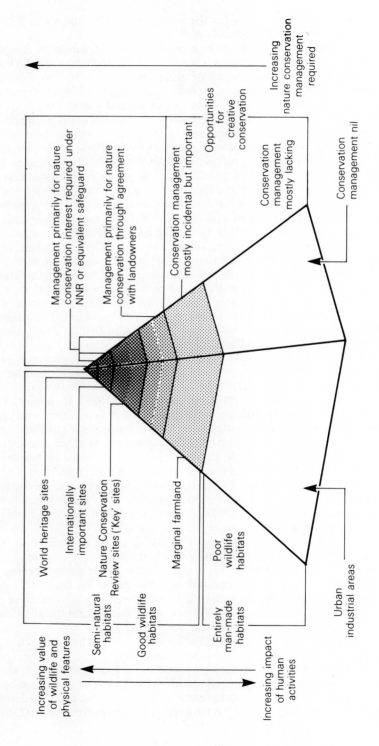

Fig 9.5 Ranked hierarchy of nature conservation sites and different management strategies. Source: Nature Conservancy Council (1984).

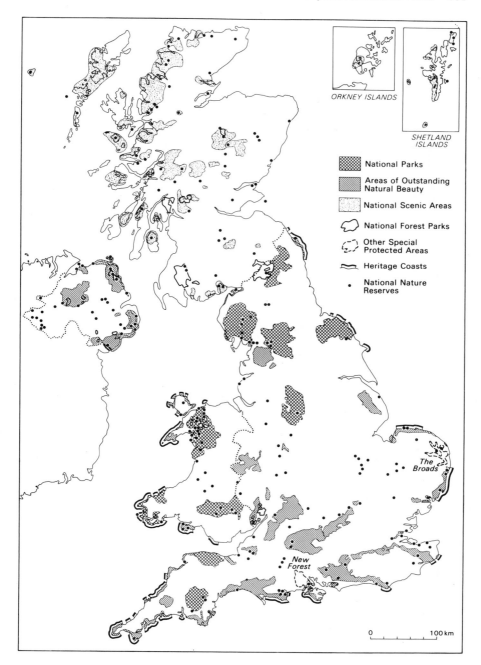

Fig 9.6 Protected landscapes and nature reserves in United Kingdom. Sources: Countryside Commission, Countryside Commission for Scotland, and Ulster Countryside Committee, Annual Reports.

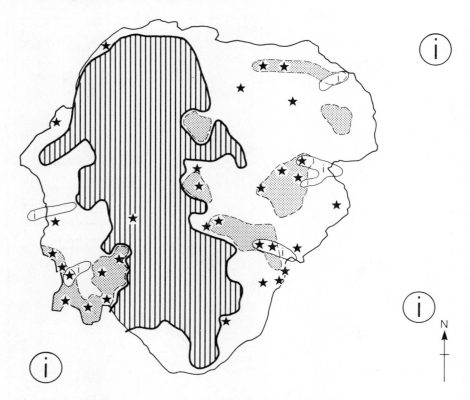

Proposed strategy for recreation and information/interpretation

★▦ Areas/sites where there is either: a) current heavy recreational use
            or: b) an apparent potential to accept
                     such use which requires further
                     investigation

▥ Areas with a low capacity for car-borne recreation

i Areas in which information and interpretation will be concentrated

Fig 9.7 An example of a National Park Management Plan: The case of Dartmoor. The Park is divided into two main areas, areas for conservation and areas for recreation. Within this major division there are however a number of detailed management policies dealing with individual sites. Source: Dartmoor NPA (1977)

1982), wildlife and habitat (Council of Europe, 1979) be passed and observed, there is also a need to divide the world into the following management areas (Forster, 1973):

1 *Natural Areas*
(a) Strict natural areas
(b) Managed natural areas

2 *Cultural Areas*
(a) Cultivated landscapes
(b) Archaeological or historical sites

(c) Wilderness areas
(d) Natural environment recreation areas

(c) Anthropological areas

in which not only conservation and recreation are reconciled but comprehensive land-use planning is practised (Nelson, 1978). Accordingly attention is now turned towards land management.

Land management and integrated rural development

Management and management agreements: the case of agriculture and conservation

Attempts to manage farmland, other than for agricultural purposes, have until recently met a good deal of opposition for a number of reasons, including the pre-eminent role given to agriculture after the food shortages of the Second World War; the view that agriculture is the natural custodian of the countryside; the fact that agriculture involves living organisms, and is therefore less amenable to the mainly negative controls used in land use planning; and finally, the political strength of the farm lobby. In summary, these have led to a weak set of conservation controls barely backed up with inadequate budgets compared to a farming industry supported by powerful legislation and buttressed by enormous cash support.

Nonetheless, the last decade has seen a growing amount of support for landscape and wildlife conservation to be given equal priority with agriculture (Davidson and Lloyd, 1978), and for agricultural agencies to accept a responsibility for conservation and amenity (Advisory Council for Agriculture and Horticulture, 1978) largely because the new agricultural habitats (discussed in Chapter 8) no longer contain the basic requirements which are essential for the survival of many wildlife species (NCC, 1984). Even in the farming industry, surveys have shown that some farmers would like to be able to reconcile commercial agriculture with conservation (Hart, 1977) although this varies with the type of farm (MAFF, 1976) and the type of farmer (Newby *et al.*, 1977) with upland and family farmers being the most favourably disposed to conservation.

Chronologically, attempts to reconcile conservation and farming really date from the 'New Agricultural Landscapes' study of the early 1970s, which helped to reveal the problem landscapes shown in Figure 9.8. The policy response to this report was based on three main approaches. First, much reliance was placed on voluntary methods and the publication and dissemination of agreed advice concerning ways in which unproductive pieces of farmland could be made better habitats, for example ponds in damp hollows, as shown in Figure 9.9 and the formation of 'Farming Wildlife Advisory Groups' (FWAGS) (Countryside Commission *et al.*, 1980). Second, the payment of grant aid to public agencies, voluntary bodies and farmers in order that they might create habitat (Countryside Commission, 1984a; 1984b), and third, the establishment of five 'New Agricultural Landscapes' projects and 10 lowland 'demonstration farms' (Lloyd, 1979) to show how conservation and farming could be reconciled both on the ground and financially (Cobham and Hockin, 1979), since the farming community is especially susceptible to the example of neighbours, as the discussion of diffusion theory in Chapter 2 demonstrated.

However it was in the uplands that the voluntary approach was first shown to be inadequate when the rate of moorland loss to agriculture, notably in Exmoor, became unacceptable. Although the Porchester report (Porchester, 1977), discussed in Chapter 8, recommended the introduction of 'Moorland Conservation Orders' over all the moorland at risk of further reclamation at the cost of a once and for all compensation payment of around £750,000, the official policy has been one of dividing the area up into two areas: the overall moorland area (Map 1)* and within this area the moorland area to be conserved for all time (Map 2) (Curtis and Walker, 1980), and then offering to any farmer wishing to reclaim land, especially in the Map 2 area, a management agreement with compensation.

Management agreements have been defined (Feist, 1978) as a voluntary agreement between one or more parties to adhere to an agreed course of action or inaction, usually in return for some sort of assistance, normally financial. The agreement usually covers one or more of the following points:

1 to refrain from an existing or proposed activity;
2 to modify an existing or proposed activity;
3 to maintain an existing activity or situation;
4 to undertake a proposed activity; and
5 to permit agencies other than the land occupier to carry out certain activities.

Management agreements can be drawn up under either the Town and Country Planning Act 1971 or the Wildlife and Countryside Act 1981. Agreements under the 1971 Act have mainly dealt with wider issues of countryside management (Probert and Hamersley, 1979), and have been used most widely in the urban fringe (Countryside Commission, 1981b). Nonetheless, they have also been used solely to reconcile agriculture and conservation, and on Exmoor, two pioneering agreements have conserved 152 hectares of moorland at an annual cost of around £6,500 over 20 years (Leonard, 1982).**

Management agreements under the 1981 Act were introduced as a major back-up to the voluntary approach, and they recognize that in certain cases conservation can only be achieved by the payment of compensation. Accordingly, the Countryside Commission will pay between 50 and 75 per cent of the cost of an agreement (Countryside Commission, 1983b) which the farming lobby hope will be a positive arrangement for conserving and enhancing desirable features on a farm (CLA *et al.*, 1984). Two types of compensation are possible, a lump sum or an annual payment (DOE, 1983) but the annual budget allocated to such payments has been set at only about £1m. As Table 9.9 shows,

---

* Under the Wildlife and Countryside Act 1981 each National Park Authority has to divide their moorland area into Map 1 and Map 2 areas. Map 1 is the overall area of moorland and Map 2 is a smaller area of moorland which should be preserved if at all possible. Exmoor was in the vanguard of this mapping exercise but all National Parks had completed the exercise by the early 1980s.
** More agreements covering larger areas have since been signed using the 1981 Act.

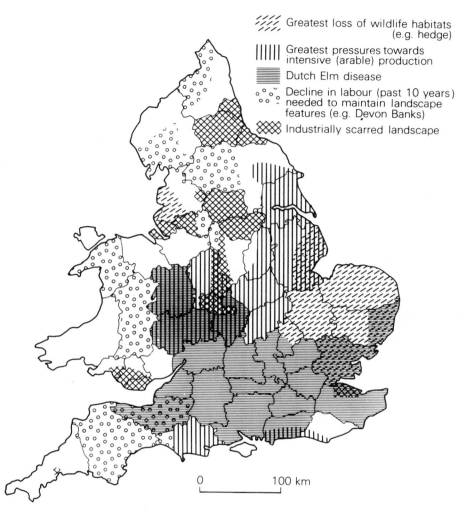

Fig 9.8 Problem rural landscapes. Landscapes can change from one or more causes, sometimes passive neglect of existing features and at other times positive forces for change like intensive arable cultivation. Source: Cobham (1984b).

this sum could easily be dwarfed by protecting one sort of site only, the Sites of Special Scientific Interest (SSSIs).

Many of these SSSIs, numbering about 4,000, have already been damaged or totally destroyed by farming operations, and although management agreements specifically for SSSIs, under the 1968 Countryside Act and the 1981 Wildlife and Countryside Act, have stemmed the tide of destruction, it is feared that many more could be lost (Barton and Buckley, 1983) and so the 1981 Act also provides for public land acquisition if the voluntary approach should fail (DOE, 1982c).

In summary then, the voluntary approach has attempted to marry conservation and farming, so that they go hand in hand in the wider landscape, and to use management agreements only in key conservation sites. However, even the

Site	Year in which work to be done	Feature to be conserved or created	Management work	Potential source of grant
1.	Year 1	Belt of trees, which screen and provide shelter for farm buildings. Also potential source of fuel.	Remove over-mature spruce trees over 2 years. Phased replanting with broadleaved species eg oak, ash and shrubs.	Countryside Comission
2.	Year 1	Belt of trees which screen and shelter farm buildings	Remove over-mature spruce trees over 2 years. Plant 3 to 5 trees eg oak, ash to break up line of buildings when viewed from north. Allow adequate space between trees to allow sun to penetrate to buildings, particularly during winter. Plant hedge along boundary between present tree belt and field to shelter buildings at ground level.	Countryside Commission Ministry of Agriculture Fisheries and Food (MAFF)
3.	Year 3	Green lane to be conserved for landscape and historical reasons.	Hedges bordering lane to be trimmed every 2 or 3 years in late winter or early spring.	
4.	Year 2	Wooded area bordering pool to be conserved as landscape feature and wildlife habitat. Source of fuel.	Over mature and some of derelict trees to be removed at intervals for use as fuel. A few of derelict trees to be left for wildlife. Some ash to be coppiced to provide source of fuel and poles for use on farm. Trees, eg oak, ask to be planted in spaces created by removal of trees, eventually for use as fuel or for sale as timber!	Countryside Commission Forestry Commission
5.	Year 5	Alders along stream edge to be retained as landscape feature.	Alders to be pollarded in phases at 15 year intervals to provide fuel.	Countryside Commission
6.	Year 2	Hedgerow trees to be retained as feature on skyline.	Three to five trees to be planted as eventual replacements for existing oak trees.	Countryside Commission
7.	Year 4	Create pond.	Pond to be dug in existing wet area adjacent to stream, which will provide wildlife habitat. May provide opportunity for wildfowl shooting.	Countryside Commission
8.	Year 3	Plant hedge.	Plant hedge to replace existing wire fence, which will reduce wind speed and provide shelter for stock, while also creating attractive landscape feature.	MAFF
9.	Year 1	Wild daffodils.	Field to be retained as pasture in order to protect wild daffodils. Field can be grazed between end of May and beginning of February each year.	Nature Conservancy Council

Brookfield Farm Conservation Scheme

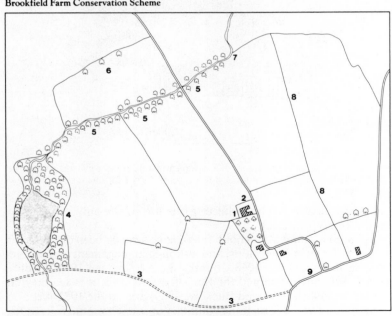

Fig 9.9 Methods for improving the wildlife habitat and landscape of a lowland farm. Note the generally low key and piecemeal changes which leave the main commercial part of the farm untouched. Source: Countryside Commission (1984a) Reproduced with the permission of the Countryside Commission.

Table 9.9: Past and possible future costs of management agreements in SSSIs

Past costs under Section 15 of the Countryside Act 1968

Year	Total number	Area (ha)	Cost in each year
Existing 1 Nov. 1973	5	118	n.a.
1973–75	9	236	n.a.
1976–77	16	816	n.a.
1978–79	27	1,500	n.a.
1979–80	43	2,228	£25,471
1980–81	70	2,577	£38,078
1981–82	90	3,032	£60,550
1982–83	111	3,112	£47,546.

Estimated costs of possible management agreements on all SSSIs

	Area (million ha)	Estimated annual compensation	Annual total
Woodland	0.3	£50	£15m
Lowland grass	0.21	£100	£21m
Peat	0.14	£30	£4.2m
Upland	0.13	£20	£2.6m
		Annual total	£42.8m
		Total over 10 years	£426m
		Total over 20 years	£850m

Source: Adams (1984).

Countryside Commission (1984c) has admitted that the future of the voluntary approach is on trial, and that if it fails, other approaches based on more draconian legislation may have to be tried. Accordingly, the last part of this section assesses the effectiveness of voluntary methods and management agreements.

Studies of 'New Agricultural Landscapes' type projects have found them to be a qualified success, in that they have encouraged the establishment of new habitat, but have mostly failed to conserve existing features (Hamilton and Woolcock, 1984), and that the acceptability of conservation proposals has been directly related to their degree of conflict with farming operations (Cobham, 1984a). The 'demonstration farms' project has achieved a greater success, both in producing integrated land use plans, and also in demonstrating to a large audience, the feasibility of marrying conservation and agriculture in a commercial farming context (Cobham, 1984b).

Nonetheless serious doubts exist. Perhaps the most fundamental query is about the cost, and as Table 9.9 shows, the small and acceptable costs of protection in the 1970s could rise to staggering amounts even when only SSSIs are taken into account. In a wider critique the MacEwens (1982) have not only echoed the limited areal nature of protection (largely restricted to SSSIs and national parks) under the 1981 Act, but have also roundly criticized the concept of compensation, which flies against the practice in all other areas of land use planning, and which they see as a 'dead end' that pours small amounts of money into a bottomless pit. Cowap (1982) also doubts whether there is even the level of goodwill needed to reach agreement, even when compensation is being offered.

Accordingly, a number of commentators have proposed a number of alternative policy options. These have included: (a) the introduction of planning controls over farming operations (Shoard, 1980) which by themselves may be too negative; (b) the extension of public ownership, but although this has been the main method used in the past in the USA (Kusler, 1980) it is probably too expensive given today's land values and (c) the use of easements (or restrictive covenants), but in line with management agreements in the UK, experience in the USA has shown that they can be very expensive (Strong, 1983).

All of these approaches may however be tackling the symptoms and not the cause, and a number of other commentators have attributed the conflict to the over-subsidization of agriculture by government support policies. Indeed, Bowers and Cheshire (1983) have argued that a reduction in farm support would markedly reduce the conflict between agriculture and conservation. However, Potter (1983) has pointed out that a reduction in support may only drive the farmer to even higher levels of production to maintain his income, and has therefore argued for a new system of support which gives as much emphasis to conservation as it does to increasing productivity. This proposal has been echoed by Wibberley (1982) who has proposed a wider role for the Ministry of Agriculture, to embrace conservation as well as farming, and the development of a multi-purpose rural land management policy. Such calls are not however entirely new nor have they as yet been too successful as the last section will demonstrate.

Multiple rural land use and integrated rural development
For a long time, many people have seen the introduction of multiple land use, via integrated rural development, as the great panacea and goal of rural planning, and it has now begun to feature strongly in textbooks comparing international experience, for example, Burnell-Held and Visser's (1984) excellent comparison of rural planning in the USA and Holland. However, integrated rural development is a confusing phrase which has no single meaning (McNab, 1984), and although it has been used to describe economic and social development only, for example the Integrated Rural Development programme for the Western Isles of Scotland, already outlined in this chapter, in this context it is defined as the development of economic, social and conservation goals.

Calls for such integrated rural development have frequently been made in various forms and by various bodies. For example, in 1976 a House of Commons Committee (Expenditure Committee HC, 1976, xviii) asked the government: 'to formulate an overall strategy for land use in the countryside', the Countryside Review Committee (CRC 1979, 24) has argued that 'conservation can no longer be viewed as a luxury but must be considered as a basic element in the management and decision taking process'; and the Department of the Environment (DOE 1977, paragraph 24) concurred with this view when it pointed out that: 'the general survival of the nation's wildlife cannot be achieved solely by site protection but depends on the wise management of the nation's land resources as a whole. Consequently, nature conservation has a place in all activities affecting rural land use and planning'. Finally, in the uplands the Countryside Commission (1984d) after a lengthy consultation process has found that all respondents were unanimous in calling for better coordination between all government activities, since on the one hand, departments pay little regard to the decisions and policies of other departments, while on the other hand, their grant aid schemes and other forms of assistance often appear either duplicatory or contradictory.

These conclusions have also been made at the international level and the

Commission of the European Communities (1984, 17) has argued that: 'major aspects of environment policy in individual countries must no longer be planned and implemented in isolation'. Perhaps the most important statement on integrated rural development has however come in the 'World Conservation Strategy' produced by the International Union for Nature Conservation (IUCN, 1980). This 'Strategy' concluded that improvements in economic development and human happiness cannot be maintained unless the conservation of living and material resources is specifically drawn into the developmental process, by making use of the concept of 'the sustainable utilization of resources'. A good example of this concept is provided by Finnish and Swiss forestry practice, where only the amount of timber equal to the annual increment may be harvested in Finland, and the total area of the forest in Switzerland cannot be reduced.

In the UK context, O'Riordan (1982) has argued that the 'Strategy' should try and embrace the three concepts of 'ecological sustainability', 'aesthetically sustainable utilization' and 'socially and economically viable rural communities' to produce a system of integrated rural resource management. However, this may be far harder to achieve than to advocate. For a start, there are still no theoretically or economically agreed methods of allocating land uses and resources between different users and areas, although a number of attempts have been made (Maxwell *et al.*, 1979; Selman, 1977). Nonetheless, as Chapter 3 has already demonstrated, both privately and publicly owned forest estates have achieved a good deal of integration on the ground, and in the Alps the long tradition of multiple land use has been extended and fairly successfully widened by the development of all-year-round ski and sports resorts (Netting, 1981).

Progress has also been made in the form of experimental management schemes in the Lake District (Countryside Commission, 1976) and Snowdonia (Countryside Commission, 1979) but although these were fairly successful in getting project officers to implement conservation and recreation schemes on individual farms, they were by and large cosmetic and treated the symptoms rather than the causes. Accordingly, the next development has been a more fundamental approach in the Peak District (Peak Park Joint Planning Board, 1984) where an 'Integrated Rural Development Experiment' has sought to not only revive declining rural communities but to sustain the environment as well by paying farmers to conserve existing features rather than to increase production (Sluman, 1984). For example, grants for the upkeep of dry-stone walls provide three functions: *environmental*, by maintaining a characteristic landscape feature; *economic*, by retaining stockproof boundaries; and *social*, by encouraging the local craft of wall building.

In the wider landscape, county councils have also tried to include conservation as part of their multiple land use policies but so far with limited success. For example, most counties have made substantial progress in producing ecological inventories, but have not as yet gone any further than this initial data collection exercise (Selman, 1982) and in the area of landscape conservation Penning-Rowsell (1983) from a survey of county councils has found first, that there is a mismatch between the decline of landscape quality and the planning response, and second, that existing planning controls can anyway have little effect on landscape changes since these are due to more powerful economic and technological forces.

In conclusion, it is clear that as yet little progress has been made in implementing integrated rural development, largely because of the entrenched and sectoral viewpoints of most of the bodies involved (NCC, 1984). Accordingly, some commentators have argued for a rearrangement of land management zones as a prelude to achieving overall integration. For example, O'Riordan (1982) suggests a division into:

(a) heritage sites;

(b) conservation zones; and

(c) agricultural and forestry landscapes

and Green (1981) a division into:

(a) field features;

(b) farm features;

(c) local reserves and public open spaces;

(d) country or, perhaps county, parks and reserves; and

(e) regional conservation areas.

However, such systems still fall into the trap of compartmentalizing areas, which is just as much a problem as compartmentalizing authorities and powers, and it is hard to see the concepts of 'sustainable utilization', and 'overall integration' coming into reality unless there is some adjustment in the fundamental attitudes of countryside planners and users, and those who must arbitrate in rural resource decisions (Davidson and Wibberley, 1977). In the meantime, rural planning and land management will seriously fail to achieve its goals, and those many rural geographers who are involved in rural planning, as either practitioners or commentators, will have to continue to argue for a more integrated approach.

10

Rural geography: some concluding questions

This book began by discussing the relationship between geography and rural geography. It is now time to ask if there is indeed a discipline of rural geography, to also enquire what are the themes and questions that have dominated the subject in recent years, and finally to query what direction the subject might take in the future.

Is there a discipline of rural geography?

The answer to this question must be a qualified no, for although this book has identified a number of common themes and approaches, these can also be identified within the wider field of geography. For example, most of the chapters in this book have revealed a common *methodological* approach, in which data collection has been followed by a period of description and explanation, which has then often led to a period of elementary model building and theory construction, and then finally, to an increasing interest in policy formulation and analysis, as well as a more behavioural approach to the subject. Another common approach can be found in the widespread use of *multivariate* statistics, but once again this is not unique to rural geography.

Only in the area of *multiple* land use and the *integration* of eclectic studies and policies can there be a partial claim that rural geographers have created a separate approach, apart from the fact that rural areas are different from towns by their very nature. In conclusion therefore, rural geography is very much a parallel discipline within geography, and is an approach that has derived much of its methodology from other branches of the subject, rather than creating its own individuality.

What are the common themes and questions addressed by contemporary rural geography?

It is not the purpose of this section to summarize the individual themes discussed in each chapter, but instead to examine the common issues that have cut across these topic headings, and in particular, the themes outlined by Cloke (1980) in Chapter 1. By far and away the most obvious of these themes, is the *information explosion*, and the subject has seen not only the new literature outlined in Table 1.1, but each year the *Countryside Planning Yearbook* notes the publication of about 250 books, reports, reviews and monographs, and another 200 or so articles, even in the more restricted field of countryside planning (Gilg, 1982).

At the same time however, the publication of the *Countryside Planning Year-*

book reflects the need to draw together what is still a very eclectic, diverse and widely spread literature as portrayed by the different approaches outlined in each chapter. There is little evidence therefore, that rural geography has become any more coherent, or that it has produced a widely accepted body of theory, and indeed the subject remains broadly *theory free*. Furthermore many of the theories date from the nineteenth century, for example, Ravenstein's laws of migration, and Von Thünen's agricultural location theory, or from the 1930s and 1940s, for example, central place theory. In newer branches of the subject, like rural recreation and landscape evaluation, explanatory rather than causal models have been produced, and even these have often been roundly criticized.

Another theme that has attracted a good deal of attention and criticism is the *definition* of rural areas. For example, the traditional definitions based on landscape and land use may no longer be acceptable in the socioeconomic field, and although a number of multivariate techniques have provided quite distinctive definitions, other approaches have pointed out that the differences between urban and rural areas are more of a continuum, for example, in the cases of deprivation and social communities.

Another area where the distinction between town and country is becoming more blurred, is in the area of counterurbanization, and critics of the rural geographical approach to this issue have pointed out that this has examined the patterns produced rather than the underlying processes. Nonetheless, one final theme is that more attention, albeit not enough, is now being paid to the *spatial dynamics* of rural areas, and how they produce the patterns that can and have been observed. Prime examples of this approach are provided by behavioural studies of farmers, and analyses of land use planning policies.

What direction might rural geography now take?

In 1980 Cloke proposed three new emphases for rural geography; the development of a *conceptual* framework, the isolation of suitable *analytical tools*, and an *integrated approach* for rural geography and planning. Taking these in turn, it can be argued that rural geography is no nearer a *conceptual* framework in 1984, than it was in 1980. One way round this would be to subdivide the subject into those areas where common concepts could be more easily applied, but this would in many ways be self-defeating. It is probably better to argue that the need for a conceptual framework is probably a passing academic fashion, and that the best concept for rural geography should be a spatial one, in which rural geography is seen as the study of extensive land uses or activities in which the friction of distance plays a major role.

Cloke's second new emphasis, was the isolation of suitable *analytical tools*, and although this issue represents nearly as big a red herring as the need for a conceptual framework, it does at the same time present more of a problem, in that the search for uniquely spatial statistical methods, since the quantitative revolution of the 1960s continues to be fruitless. This is a problem, because the definition of rural geography employed above specifically refers to the friction of distance in rural areas, and therefore the lack of spatially based statistical methods is a major weakness in the rural geographer's armoury.

Nonetheless, there has been one area where rural geographers have made great advances, notably in Cloke's third area, the *integrated approach* to rural geography and planning. For example, rural geographers have not only assessed the effect of rural planning policies, but have helped to evaluate and rewrite them, and geography graduates are now in positions of power in many environmental agencies, for example, Derek Phillips is the second director of the Countryside Commission to be a geography graduate, and many National Park Officers are also geography graduates. Although the Thatcher and Reagan years have tended to slow this advance down, the verities of the rural environment will not go away, and the message of the sustainable use of resources contained in the World Conservation Strategy will eventually have to be understood and acted upon. In many ways therefore, the diversions of the past 20 years, into methodological introspection and statistical debate, have been in the long term rather arid years. There are now however, encouraging signs that the lessons of these years can now be harnessed, and that a more robust, academically sound and applied rural geography can take on the environmental challenges of the 1980s and 1990s.

In conclusion therefore, the future for rural geography should be an applied one, where it integrates its own research, relates this to the real behavioural world and to policy formulation, and thus attempts to produce a rural environment that is not only physically attractive but also a lively and prosperous place to live.

Bibliography

ACC 1979: *Rural deprivation*. London: ACC.

Adams, W.M. 1984: *Implementing the Act: A study of habitat protection under Section II of the Wildlife and Countryside Act 1981*. Godalming: WWF.

ADC 1983: *Economic developments by district councils*. London: ADC.

Advisory Council for Agriculture and Horticulture 1978: *Agriculture and the country-side*. London: MAFF.

Agricultural Land Service 1966: *Agricultural land classification*. Pinner: MAFF.

Aitchison, J.W. and Aubrey, P. 1982: Part-time farming in Wales: A typological study. *Transactions, IBG* **7**, 88–97.

Allaby, M. 1983: *The changing uplands*, Cheltenham: Countryside Commission.

Allan, J.A. 1980: Remote sensing in land and land use studies. *Geography* **65**, 35–43.

Ambrose, P. 1974: *The quiet revolution: Social change in a Sussex village 1871–1971*, London: Chatto & Windus.

Anderson, J.R., Dillon, J.L. and Hardaker, J.B. 1977: *Agricultural decision analysis*, Ames: Iowa State University Press.

Anderson, K.E. 1975: An agricultural classification of England and Wales. *Tijdschrift voor Economische en Sociale Geografie* **66**, 148–58.

Anderson, M.A. 1975: Land planning implications of increased food supplies. *The Planner* **61**, 381–3.

——1977: A comparison of figures for the land use structure of England and Wales in the 1960s. *Area* **9**, 43–45.

——1980: The land pattern of Areas of Outstanding Natural Beauty in England and Wales. *Landscape Planning* **7**, 1–22.

——1981a: Planning policies and development control in the Sussex Downs AONB, *Town Planning Review* **52**, 5–25.

——1981b: The proposed High Weald Area of Outstanding Natural Beauty. *Landscape Research* **6**, 6–10.

——1984: Complete urban containment: A reasonable proposition? *Area* **16**, 25–31.

Anderson, P. and Yalden, D.W. 1981: Increased sheep numbers and the loss of heather moorland in the Peak District, England. *Biological Conservation* **20**, 195–213.

Andreae, B. 1981: *Farming development and space. World agricultural geography*, New York: Walter de Gruyter.

Anon 1982: Why America and Europe don't see eye to eye over agriculture. *Europe* August, 26–7.

Appleton, J.M. 1975: *The experience of landscape*. London: John Wiley.

Arensberg, C.M. and Kimball, S.T. 1940: *Family and community in Ireland*. London: Peter Smith.

Armstrong, A. 1982: The failure of planning in the Highlands. *Town and Country Planning* **51**, 99–101.

Armstrong, P. 1975: *The changing landscape*, Lavenham: Terence Dalton.

Ashby, A.W. 1939: The effects of urban growth on the countryside. *Sociological Review* **31**, 345–69.

Ashby, P.; Birch, G. and Haslett, M. 1975: Second homes in North Wales. *Town Planning Review* **46**, 323–33.

Ashton, J. and Cracknell, B.E. 1961: Agricultural holdings and farm business structure in England and Wales. *Journal of Agricultural Economics* **14**, 472–506.

Askew, I. 1983: The location of service facilities in rural areas: A model for generating and evaluating alternative solutions. *Regional Studies* **17**, 305–14.

Baker, S. 1974: *Milk to market: Forty years of milk marketing*. London: Heinemann.

Balchin, W.G.V. 1981: Land use mapping in the 1980s. *Countryside Planning Year Book* **2**, 188–9.

Ball, R.M. 1980: The use and definition of travel to work areas in Great Britain: Some problems. *Regional Studies* **14**, 125–39.

Banister, D. 1980: Transport mobility in Interurban areas: A case study approach in South Oxfordshire. *Regional Studies* **14**, 285–96.

——1983: Transport and accessibility. In Pacione 1983, 130–48.

Barker, D. 1977: The paracme of innovations: The neglected aftermath of diffusion or a wave goodbye to an idea. *Area* **9**, 259–65.

Barnes, F.A. 1958: The evolution of the salient pattern of milk production and distribution in England and Wales. *Transactions, IBG* **25**, 167–95.

Barrass, R. 1979: The first ten years of English structure planning. *Planning Outlook* **22**, 19–23.

Barrett, E.C. and Curtis, L.F. 1976: *Introduction to environmental remote sensing*. London: Chapman & Hall.

Barton, P.M. and Buckley, G.P. 1983: The status and protection of notified Sites of Special Scientific Interest in south-east England. *Biological Conservation* **27**, 213–42.

Baxter, M. and Ewing, G. 1981: Models of recreational trip distribution. *Regional Studies* **15**, 327–44.

Beale, C.L. 1983: The population turnaround in rural small town America. In Browne, W. and Hadwiger, D. (eds.) *Rural policy problems* (Lexington: Lexington Books) 47–59.

Bealer, R.C.: Willits, F.K. and Knulesky, W.P. 1965: The meaning of rurality in American society. Some implications of alternative definitions. *Rural Sociology* **30**, 255–66.

Belding, R. 1981: A test of the von Thünen locational model of agricultural land use with accountancy data from the European Economic Community. *Transactions, IBG, New Series* **6**, 176–87.

Bell, T., Ceiber, S. and Rushton, G. 1974: Clustering of services in central places. *AAAG* **64**, 214–25.

Berry, D. and Plaut, T. 1978: Retaining agricultural activities under urban pressures: A review of land use conflicts and policies. *Policy Sciences* **9**, 153–78.

Best, R.H. 1976: The changing land use structure of Britain. *Town and Country Planning* **44**, 171–6.

——1977: Agricultural land loss: Myth or reality? *The Planner* **63**, 15–16.

——1979: Land use structure and change in the EEC. *Town Planning Review* **50**, 395–411.

——1981: Land use and living space. London: Methuen.

Best, R.H. and Champion, A.G. 1970: Regional conversions of agricultural land to urban use in England and Wales, 1945–67. *Transactions, IBG* **49**, 15–31.

Best, R.H.; Jones, A.R. and Rogers, A.W. 1974: The density size rule. *Urban Studies* **11**, 201–8.

Best, R.H. and Rogers, A.W. 1973: *The urban countryside*. London: Faber.

Bielckus, C.L.; Rogers, A.W. and Wibberley, G.P. 1972: *Second homes in England and Wales*. Ashford: Countryside Planning Unit.

Birch, F. 1979: Leisure patterns 1973 and 1977. *Population Trends* **19**, 2–8.

Blacksell, M. 1979: Landscape protection and development control: An appraisal of

planning in rural areas in England and Wales. *Geoforum* **10**, 267–74.
——1981: Analysing land use change. *Geography* **66**, 116–23.
——1983: Leisure recreation and environment. In Johnston and Doornkamp 1983, 309–26.
Blacksell, M. and Gilg, A.W. 1975: Landscape evaluation in practice: The case of south-east Devon. *Transactions, IBG* **66**, 135–40.
——1977: Planning control in an Area of Outstanding Natural Beauty. *Social and Economic Administration* **11**, 206–15.
——1981: *The countryside: Planning and change*. London: George Allen & Unwin.
Blake, J. 1979: Job prospects. *Town and Country Planning* **47**, 7–10.
Blunden, J. 1975: *The mineral resources of Britain*. London: Hutchinson.
Boddington, M.A.B. 1978: *The classification of agricultural land in England and Wales*. Oxford: Rural Planning Services.
——1984: Finding a new role for the countryside. *Town and Country Planning* **53**, 170–2.
Bollom, C. 1978: *Attitudes and second homes in rural Wales*. Cardiff: University of Wales Press.
Bowers, J.K. and Cheshire, P. 1983: *Agriculture, the countryside and land use*. London: Methuen.
Bowler, I.R. (ed.) 1975: *Register of research in rural geography*, Leicester: Rural Geography Study Group.
——1976a: Spatial responses to agricultural subsidies in England and Wales. *Area* **8**, 225–9.
——1976b: Regional agricultural policies: Experience in the United Kingdom. *Economic Geography* **52**, 267–80.
——1976c: The adoption of grant aid in agriculture. *Transactions, IBG, New Series* **1**, 143–58.
——1979: *Government and agriculture: A spatial perspective*. London: Longmans.
——1981: Regional specialisation in the agricultural industry. *Journal of Agricultural Economics* **32**, 43–55.
——1983a: The agricultural pattern. In Johnston and Doornkamp 1983, 75–104.
——1983b: Structural change in agriculture. In Pacione 1983, 46–73.
——1984: Agricultural geography. *Progress in Human Geography* **8**, 255–62.
Bowman, J.C.; Robbins, C.J. and Doyle, C.J. 1976: The case for an agricultural strategy for the United Kingdom. *Journal of RASE* **137**, 27–33.
Boyer, J.C. 1980: Residence Secondaires et 'Rurbanisation' en region Parisienne. *Tijdschrift voor Economische en Sociale Geografie* **71**, 78–87.
Bracey, H.E. 1956: A rural component of centrality applied to six southern counties in the United Kingdom, *Economic Geography* **32**, 38–50.
——1958: Some aspects of rural depopulation in the United Kingdom. *Rural Sociology* **23**, 385–91.
Bracken, I. 1980: Structure plans: Alterations and submissions. *The Planner* **66**, 130–2.
Bradshaw, T.K. and Blakeley, E.J. 1979: *Rural communities in an advanced industrial society*. New York: Praeger.
——1983: The changing nature of rural America. In Browne, W. and Hadwiger, D. (eds.) *Rural policy problems* (Lexington: Lexington Books) 3–18.
Brant, J. 1984: Patterns of migration from the 1981 Census. *Population Trends* **35**, 23–30.
Bray, C. 1981: *New villages: Case study no. 1: New Ash Green*. Oxford: Oxford Polytechnic, Town Planning Department.
Briggs, D.J. and France, J. 1980: Landscape evaluation: A comparative study. *Journal of Environmental Management* **10**, 263–75.
Britton, D.K. 1968: Comment on 'The geography of agriculture' by J.T. Coppock. *Journal of Agricultural Economics* **29**, 153–75.

Britton, D.K. and Hill, B. 1975: *Size and efficiency in farming*. Farnborough: Saxon House.

Broady, M. 1980: Mid-Wales: A classic case of rural self-help. *The Planner* **66**, 94–6.

Brotherton, D.I. 1975: The development and management of country parks in England and Wales. *Biological Conservation* **7**, 171–84.

——1982a: Visitor frequency and fidelity as indicators of rural recreation provision. *Journal of Environmental Management* **15**, 101–7.

——1982b: Development pressures and control in the National Parks. *Town Planning Review* **53**, 439–59.

——1982c: National parks in Great Britain and the achievement of nature conservation purposes. *Biological Conservation* **22**, 85–100.

——1983: Determinants of landscape change: The case of afforestation in the National Parks of England and Wales. *Landscape Planning* **9**, 193–207.

Brown, D.L. and Wardwell, J.M. (eds.) 1980: *New directions in urban-rural migration: The population turnaround in North America*. New York: Academic Press.

Browne, W. and Hadwiger, D. (eds.) 1983: *Rural policy problems*. Lexington: Lexington Books.

Bruton, M. 1983: Local plans, local planning and development plan schemes in England 1974–82. *Town Planning Review* **54**, 4–23.

Bryden, J. 1981: Appraising a regional development programme: The case of the Scottish Highlands and Islands. *European Review of Agricultural Economics* **8**, 475–97.

Bruton, M. 1983: Local plans, local planning and development plan schemes in England 1974–82. *Town Planning Review* **54**, 4–23.

BTA 1983: *Digest of tourist statistics no.* 11. London: BTA.

Bunce, M. 1982: *Rural settlement in an urban world*, London: Croom Helm.

Burnell-Held, R. and Visser, D.W. 1984: *Rural land uses and planning: A comparative study of the Netherlands and the United States*. Amsterdam: Elsevier.

Butler, J.E. and Fuguitt, G.V. 1970: Small town population change and distance from larger towns. *Rural Sociology* **35**, 396–409.

Buttel, F.H. 1983. Farm structure and rural development. In Browne and Hadwiger 1983, 3–18.

Buttel, F.H. and Flinn, W.L. 1977: Conceptions of rural life and environmental concern. *Rural Sociology* **42**, 544–55.

Buyhoff, G.J.; Wellman, J.D.; Koch, N.E.; Gauthier, L. and Hultman, S. 1983: Landscape preference metrics: An international comparison. *Journal of Environmental Management* **16**, 181–90.

Campbell, D. 1977: Landscape design considerations in large-scale forestry operations in Great Britain, *Planning Outlook* **20**, 6–14.

Campbell, R.R. and Johnson, D.M. 1976: Propositions on counter-stream migration. *Rural Sociology* **41**, 127–45.

Carlson, A.A. 1977: On the possibility of quantifying scenic beauty. *Landscape Planning* **4**, 131–72.

Carpenter, E.H. 1977: The potential for population dispersal: A closer look at residential locational preferences. *Rural Sociology* **42**, 52–370.

Carroll, M.R. 1978: *Multiple use of woodlands*. Cambridge: Department of Land Economy.

Carruthers, A.O.; Crosbie, A.J. and Waugh, T.C. 1984: Mapping the 1981 population census of Scotland. In Jones, H. (ed.) *Population change in contemporary Scotland* (Norwich: Geo Books), 11–22.

Carter, C.J. 1981: Skiing versus conservation. *The Planner* **67**, 134–5.

CAS 1980: *A forestry strategy for the UK*. Reading: CAS.

Champion, A.G. 1973: Population trends in England and Wales. *Town and Country Planning* **41**, 504–9.

——1976: Evolving patterns of population distribution in England and Wales 1951–71.

Transactions, IBG, New Series 1, 401–20.
——1981: Population trends in rural Britain. *Population Trends* **26**, 20–3.
——1983: Land use and competition. In Pacione 1983, 21–45.
Chapius, R. 1973: De l'espace rural à l'espace urbain: Problèmes de typologie. *Etudes rurale* **49–50**, 122–36.
Chapman, K. 1979: *People, pattern and process*. London: Edward Arnold.
Cherry, G. (ed.) 1976: *Rural planning problems*. London: Leonard Hill.
Chisholm, M. 1962: *Rural settlement and land use*. London: Hutchinson.
——1964: Problems in the classification and use of farming-type regions. *Transactions, IBG* **35**, 91–103.
Christaller, W. 1966: *Central places in southern Germany*. Translated from the 1933 publication 'Die zentralen Orte Suddeutschlands' by C.W. Baskin. New Jersey: Prentice Hall.
Church, B.M.; Boyd, D.A.; Evans, J.A.; and Sadler, J.I. 1968: A type of farming map based on agricultural census data. *Outlook on Agriculture* **5**, 191–6.
CLA, NFU and RICS 1984: *Management agreements in the countryside*, London: Surveyors Publications.
Clark, C. 1975: The stability of village populations. *Urban Studies* **12**, 109–11.
Clark, D. 1980: Can voluntary initiatives sustain rural services at risk? *The Planner* **66**, 100–1.
——1984: Rural Council house building slumps. *Rural Viewpoint* **5**, 9.
Clark, D. and Unwin, K. 1981: Telecommunications and travel: Potential impact in rural areas. *Regional Studies* **15**, 47–56.
Clark, G. (ed.) 1978 and 1981: *A register of research in rural geography*. Lancaster: Rural Geography Study Group.
——1982a: Developments in rural geography. *Area* **14**, 249–54.
——1982b: Housing policy in the Lake District. *Transactions, IBG New Series* **7**, 59–70.
——1982c: *Housing and planning in the countryside*. London: John Wiley.
——1984: The meaning of agricultural regions. *Scottish Geographical Magazine* **100**, 34–44.
Clark, G. and Gordon, D.S. 1980: Sampling for farm studies in Geography. *Geography* **65**, 101–6.
Clark, G.; Knowles, D.J. and Phillips, H.L. 1983: The accuracy of the agricultural census, *Geography* **68**, 115–20.
Clawson, M. 1959: *Methods for measuring the demand for and value of outdoor recreation*. Washington: Resources for the future.
——1975a: *Forests, for whom and for what?* Baltimore: Johns Hopkins University Press.
——1975b: Economic and social conflicts in land use planning. *Natural Resources Journal* **15**, 473–90.
Cloke, P. 1977a: In defence of key settlement policies. *Village* **32**, 7–11.
——1977b: An index of rurality for England and Wales. *Regional Studies* **11**, 31–46.
——1978: Changing patterns of urbanisation in rural areas of England and Wales 1961–1971. *Regional Studies* **12**, 603–17.
——1979: *Key settlements in rural areas*. London: Methuen.
——1980: New emphases for applied rural geography. *Progress in Human Geography* **4**, 182–217.
——1982a: Rethinking rural settlement planning. *Tijdschrift voor Economische en Sociale Geografie* **73**, 280–5.
——1982b: Planners' attitudes to resource concentration and dispersal in rural areas. *Planning Outlook* **25**, 16–21.
——1983: *An introduction to rural settlement planning*. London: Methuen.
——(ed.) 1985. *Journal of Rural Studies*. Oxford: Pergamon Press.
Cloke, P. and Hanrahan, P. 1984: Policy and implementation in rural planning. *Geoforum* **15**, 261–9.

Cloke P. and Park, C.C. 1980: Deprivation, resources and planning: Some implications for applied rural geography. *Geoforum* **11**, 57–61.
——1982: Country Parks in National Parks: A case study of Craig-y-Nos in the Brecon Beacons, Wales. *Journal of Environmental Management* **12**, 173–85.
Cloke, P. and Shaw, D.P. 1983: Rural settlement policies in structure plans. *Town Planning Review* **54**, 338–54.
Clout, H.D. 1972: *Rural geography: An introductory survey*. Oxford: Pergamon Press.
——1977: Rural settlements. *Progress in Human Geography* **1**, 475–80.
Cmnd 6020, 1975: *Food from our own resources*. London: HMSO.
Cmnd 7458, 1979: *Farming and the Nation*. London: HMSO.
Cmnd 8804, 1983: *Annual Review of Agriculture 1983*. London: HMSO.
Cmnd 8963, 1983: *Intervention Board for Agricultural Produce: Report for the Calendar Year 1982*. London: HMSO.
Cmnd 9111, 1983. *Regional industrial development*. London: HMSO.
Coates, V.T. 1977: The future of rural small towns: Are they obsolete in post-industrial society? *Habitat* **2**, 247–58.
Cobham, R. 1984a: *Landscape and wildlife conservation on farms*. Perth.
——1984b: *Demonstration farms: CCP 170*. Cheltenham: Countryside Commission.
Cobham, R. and Hockin, R, 1979: The planning and managing of agricultural landscapes to satisfy commercial and conservation requirements. *Landscape Research* **4**, 14–15.
Coleman, A. 1969: A geographical model for land use analysis. *Geography* **54**, 43–55.
——1976: Is planning really necessary? *Geographical Journal* **142**, 411–37.
——1977: Land use planning: success or failure? *Architects Journal* **165**, 93–134.
——1980: The place of forestry in a viable land use strategy. *Quarterly Journal of Forestry* **74**, 20–9.
Coleman, A.; Isbell, J. and Sinclair, G. 1974: The comparative statics approach to British land use trends. *Cartographic Journal* **11**, 34–41.
Coleman, R. 1981: Footpath erosion in the English Lake District. *Applied Geography* **1**, 121–31.
Colenutt, R.J. 1969: Modelling travel patterns of day visitors to the countryside. *Area* **2**, 43–7.
Collins, N.J. 1972: The Cotswold water park. *Town and Country Planning* **40**, 438–40.
Commins, P. 1978: Socio-economic adjustments to rural depopulation. *Regional Studies* **12**, 79–94.
Commins, P. and Drudy, P.J. 1980: *Problem rural regions*. London: Regional Studies Association.
Commission of the EC 1977: *Importance and functioning of the European Agricultural Guidance and Guarantee Fund*. Brussels: The Commission.
——1981: *Factors influencing ownership tenancy mobility and use of farmland in the United Kingdom*. London: HMSO.
——1984: *Ten years of Community Environment Policy*. London: HMSO.
Compton, P.A. 1983: The changing population. In Johnston and Doornkamp 1983, 37–74.
Connell, J. 1978: *The end of tradition: country life in central Surrey*. London: Routledge & Kegan Paul.
Conservation Foundation 1972: *National Parks for the future*. Wahsington DC: The Foundation.
Cooke, P. 1980: Discretionary intervention and the Welsh Development Agency. *Area* **12**, 269–77.
Coppock, J.T. 1960a; Crops and livestock changes in the Chilterns 1931–51. *Transactions, IBG* **28**, 179–98.
——1960b: The parish as a geographical/statistical unit, *Tijdschrift voor Economische en Sociale Geografie* **51**, 371–26.
——1964a: *An agricultural atlas of England and Wales*. London: Faber.

——1964b: Crop, livestock and enterprise combinations in England and Wales. *Economic Geography* **40**, 65–81.

——1970: Land use data in the service of conservation. *Geographical Journal* **136**, 190–9.

——1971: *An agricultural geography of Great Britain*, London: G. Bell & Sons.

——1976a: *An agricultural atlas of England and Wales, 2nd edn.* London: Faber.

——1976b: *An agricultural atlas of Scotland.* Edinburgh: John Donald.

——(ed.) 1977: *Second homes. Curse or blessing?* Oxford: Pergamon Press.

——1978: *Land use: Review of United Kingdom statistical sources.* Oxford: Pergamon Press.

Coppock, J.T. and Duffield, B.S., 1975: *Recreation in the countryside.* London: Macmillan.

Coppock, J.T. and Sewell, W.R.D. 1975: Resource Management and public policy. *Scottish Geographical Magazine* **91**, 4–11.

Coughlin, N. 1980: Farming on the urban fringe. *Environment* **22**, 33–9.

Council of Europe 1979: *Explanatory report concerning the convention on the conservation of European wildlife and natural habitats.* Strasbourg: The Council

Countryside Commission 1970: *The demand for outdoor recreaion in the countryside.* Cheltenham: Countryside Commission.

——1976: *The Lake District upland management experiment: CCP93.* Cheltenham: Countryside Commission.

——1979: *The Snowdonia upland management experiment: CCP122.* Cheltenham: Countryside Commission.

——1980: *Trends in tourism and recreation:CCP134.* Cheltenham: Countryside Commission.

——1981a: *National Parks: A study of rural economics: CCP144,* Cheltenham: Countryside Commission.

——1981b: *Countryside management in the urban fringe: CCP136.* Cheltenham: Countryside Commission.

——1982: *Participation in informal countryside recreation: CCP152.* Cheltenham: Countryside Commission.

——1983a: *Areas of Outstanding Natural Beauty: A policy statement: CCP157.* Cheltenham: Countryside Commission.

——1983b: *Management agreements: Policy statement and grants: CCP156.* Cheltenham: Countryside Commission.

——1984a: *Conservation grants for local authorities, public bodies and voluntary organisations: CCP172.* Cheltenham: Countryside Commission.

——1984b: *Conservation grants for farmers and landowners: CCP171.* Cheltenham: Countryside Commission.

——1984c: *Agricultural landscapes: A policy statement: CCP173,* Cheltenham: Countryside Commission.

——1984d: *A better future for the Uplands: CCP162.* Cheltenham: Countryside Commission.

Countryside Commission and Countryside Commission for Scotland 1975: *Guide to countryside interpretation: 2 parts, 1: Principles and 2: Media and facilities.* London: HMSO.

Countryside Commission, NCC, ADAS, Forestry Commission, NFU and CLA 1980: *Countryside conservation handbook.* Cheltenham: Countryside Commission.

Coventry, Solihull, and Warwickshire Councils 1971: *A strategy for the sub-region. Supplementary Report 5: Countryside.* Coventry: The Councils.

Cowap, C. 1982: *Management agreements in rural planning.* Gloucester: Gloucestershire College of Arts.

Cracknell, B. 1967: Accessibility to the countryside as a factor in planning for leisure. *Regional Studies* **1**, 147–61.

Craig, J. 1972: Population potential and population density. *Area* **4**, 10–12.

——1974: How arbitrary is population potential? *Area* **6**, 44–6.

——1979: Population density: Changes and patterns *Population Trends* **17**, 12–16.

Craig, J. and Frosztega, J. 1976: The distribution of population in Great Britain by ward and parish density 1931, 1951 and 1961. *Area* **8**, 187–90.

CRC 1976: *The countryside: Problems and policies.* London: HMSO.

——1977a: *Rural communities.* London: HMSO.

——1977b: *Leisure and the countryside.* London: HMSO.

——1978: *Food production in the countryside.* London: HMSO.

——1979: *Conservation and the countryside heritage.* London: HMSO.

Cribier, F. 1973: Les rèsidences secondaires des citadins dan les campagnes francåaises. *Etudes Rurale* **49–50**, 181–204.

Cromley, R.G. 1982: The von Thünen model and environmental uncertainty. *AAAG* 72, 404–10.

Cross, D.T. and Bristow, M.R. (eds.) 1983: *English structure planning.* London: Pion.

Cruickshank, A. 1981: USA Census: A note. *Scottish Geographical Magazine* **97**, 175–82.

Cullingford, D. and Openshaw, S. 1982: Identifying areas of rural deprivation using social area analysis. *Regional Studies* **16**, 409–18.

Cullingworth, J.B. 1982: *Town and country planning in Britain*, 8th edition. London: George Allen & Unwin.

Curry, N. 1980: *A review of cost-benefit techniques in rural recreation planning.* Gloucester. Gloucestershire College of Art and Design.

——(ed.) 1981: *Rural settlement policy and economics.* Gloucester: Gloucestershire College of Arts.

Curtis, L. and Walker, A.J. 1980: Exmoor: A problem of landscape planning and management. *Landscape Design* **130**, 7–13.

Daft, L.M. 1983: The rural poor. In Brown and Hadwiger 1983, 73–79.

DART 1977: *Second homes in Scotland.* Totnes: DART.

Dartmoor NPA 1977: *Dartmoor National Park Plan.* Exeter. Dartmoor NPA.

Davidson, J. and Lloyd, R. (eds.), 1978: *Conservation and agriculture.* Chichester: John Wiley.

Davidson, J. and Wibberley, G. 1977: *Planning and the rural environment.* Oxford: Pergamon.

Davies, E.T. 1983: *The role of farm tourism in the less favoured areas of England and Wales 1981.* Exter: Agricultural Economics Unit.

Davies, M.C. 1970: The rural community in Central Wales: A study in social geography. *Sociologia Ruralis* **10**, 143–61.

DBRW 1982: *Annual Report 1981–82.* Newtown: DBRW.

Dean, C. 1983: Keeping the wheels turning. *Town and Country Planning* **52**, 272–3.

Dean, K.; Brown, B.; Perry, R. and Shaw, D. 1984a; The conceptualisation of counterurbanisation. *Area* **16**, 9–14.

Dean, K.; Shaw, D.; Brown, B.; Perry, R. and Thorneycroft, W. 1984b: Counterurbanisation and the characteristics of persons migrating to West Cornwall. *Geoforum* **15**, 177–90.

Deane, G.C. 1980: Preliminary evaluation of SEASAT-A SAR data for land use mapping. *Geographical Journal* **146**, 408–18.

Dearden, P. 1980: A statistical technique for the evaluation of the visual quality of the landscape for land use planning purposes. *Journal of Environmental Management* **10**, 51–70.

——1981: Landscape evaluation: The case for a multi-dimensional approach. *Journal of Environmental Management* **13**, 95–105.

Dearden, P. and Rosenblood, L. 1980: Some observations on multivariate techniques in landscape evaluation. *Regional Studies* **13**, 99–110.

De Jong, G.F. and Humphrey, C.R. 1976: Selected characteristics of metropolitan to non metropolitan area migrants in Pennsylvania. *Rural Sociology* **41**, 526–38.

Denham, C. 1984: Urban Britain. *Population Trends* **36**, 10–18.

Denman, R. 1978: *Recreation and tourism on farms, crofts and estates.* Edinburgh: STB.

Dennier, A. 1980: National park plans. *Countryside Planning Yearbook* 1, 49–66.
DES, 1965: *Report of the Land Use Study Group. Forestry, agriculture and the multiple use of rural land*. London: HMSO.
De Vane, R. 1975: *Second home ownership: A case study*. Cardiff: University of Wales Press.
Development Commission 1982: *Fortieth Report. 1981–82: House of Commons Paper 519 (81–82)*. London: HMSO.
Devis, T. 1984: Population movements measured by the NHS central register. *Population Trends* 36, 18–24.
Dickenson, R.E. 1947: *City, region and regionalism*. London: Routledge & Kegan Paul.
Dickinson, G.C. and Shaw, M.G. 1978: The collection of national land use statistics in Great Britain: A critique. *Environment and Planning A* 10, 295–303.
Dillard, J.E. 1983: State land use policies and rural America. In Browne and Hadwiger 1983, 135–50.
Dinse, J. 1983: Towards the public interest: Redefining rural policy needs. In Browne and Hadwiger 1983, 135–50.
DOE 1972: *Sinews for survival*. London: HMSO.
——1974: *Report of the National Park Policies Review Committee*. London: HMSO.
——1975: *National land use classification*. London: HMSO.
——1976. *Report of the National Parks Policy Review Committee: Circular 4/76*. London: HMSO.
——1977: *Nature conservation and planning: Circular 108/77*. London: HMSO.
——1978a: *English House Condition Survey*. London: HMSO.
——1978b: *Developed areas 1969: A survey of England and Wales*. London: DOE.
——1979: *Memorandum on structure and local plans: Circular 4/79*. London: HMSO.
——1980: *Development control: Policy and practice: Circular 22/80*. London: HMSO.
——1981: *Coal and the environment: A report by the Commission on Energy and the Environment*. London: HMSO.
——1982a: *Vale of Belvoir Coal Field Inquiry: Report*, London: HMSO.
——1982b: *English House Condition Survey*. London: HMSO.
——1982c: *Code of guidance for sites of Special Scientific Interest*. London: HMSO.
——1983: *Wildlife and Countryside Act 1981: Financial guidelines for management agreements*. London: HMSO.
Doren, C.S.; Priddle, G. and Lewis, J. (eds.) 1979: *Land and leisure: Concepts and methods in outdoor recreation*. London: Methuen.
Downing, P. and Dower, M. 1972: *Second homes in England and Wales*. Cheltenham: Countryside Commission.
Doyle, C. 1977: The land use jigsaw: Land for forestry in the future. *Quarterly Journal of Forestry* 72, 129–39.
Doyle, C. and Tranter, R. 1978: In search of vision: Rural land use problems and policies. *Built Environment* 4, 289–310.
Drudy, P.J. 1978: Depopulation in a prosperous agricultural sub-region. *Regional Studies* 12, 49–60.
Drudy, P.J. and Drudy, S.M. 1979: Population mobility and labour supply in rural regions: North Norfolk and the Galway Gaeltacht. *Regional Studies* 13, 91–9.
Drudy, S.M. 1975: The occupational aspirations of rural school leavers. *Social Studies* 4, 230–41.
Drudy, S.M. and Drudy, P.J. 1977: Problems of development in remote rural regions. *Anglo-Irish Studies* 3, 56–70.
Duffy, P.J. 1983: Rural settlement change in the republic of Ireland: A preliminary discussion. *Geoforum* 14, 185–91.
Dunn, M.; Rawson, M. and Rogers, A. 1981: *Rural housing: Competition and choice*. London: George Allen & Unwin.
Edwards, A. and Rogers, A. (eds.) 1974: *Agricultural resources: An introduction to the farming industry of the United Kingdom*. London: Faber & Faber.
Edwards, C.J.W. 1980: Complexity and change in farm production systems: A

Somerset case study. *Transactions, IBG New Series* **5**, 45–52.

Edwards, J.A. 1971: The viability of lower size-order settlements in rural areas: The case of north-east England. *Sociologia Ruralis* **11**, 246–77.

——1973: Rural migration in England and Wales. *The Planner* **59**, 450–3.

Edwards, S.L. and Dennis, S.J. 1976: Long distance day tripping in Great Britain. *Journal of Transport Economics and Policy* **10**, 237–56.

Elson, M. 1981: Structure plan policies for pressured rural areas. *Countryside Planning Yearbook* **2**, 49–70.

Engledow, F. and Amey, L. (eds.) 1980: *Britain's future in farming*. Hemel Hempstead: Geographical Publications.

Errington, A. 1980: Occupational classification in British Agriculture. *Journal of Agricultural Economics* **31**, 73–80.

Es, V.J.C. and Brown, J.E. 1974: The rural – urban variable once more: Some individual level observaions. *Rural Sociology* **39**, 373–91.

Estall, R. 1982: Planning in Appalachia: an examination of the Appalachian regional development programme and its implications for the future of the American Regional Planning Commissions. *Transactions, IBG, New Series* **7**, 35–58.

European Communities 1980: *Yearbook of agricultural statistics 1975–1978*. London: HMSO.

——1984: *Agriculture in the United States and the European Community*. London: HMSO.

Everett, R.D. 1978: The wildlife preferences shown by countryside visitors. *Biological Conservation* **14**, 75–84.

——1979: The monetary value of the recreational benefits of wildlife. *Journal of Environmental Management* **8**, 203–13.

Expenditure Committee, HC 1976: *National Parks and the countryside: House of Commons Paper 433 (75–76)*. London: HMSO.

Feist, M. 1978: *A study of management agreements: CCP 114*. Cheltenham: Countryside Commission.

Ferguson, M.J. and Munton, R.J.C. 1979: Informal recreation sites in London's Green Belt. *Area* **11**, 196–205.

Fielding, A. 1982: Counterurbanisation in Western Europe. *Progress in Planning* **17**, 1–51.

Fisher, D.W., Lewis, J. and Priddle, G. (eds.) 1974: *Land and leisure: Concepts and methods in outdoor recreation*. Chicago: Maaroufa Press.

Fisher, J.S. and Mitchelson, R.L. 1981: Forces of change in the American settlement pattern. *Geographical Review* **71**, 298–310.

Fitton, M. 1979: Countryside recreation: the problems of opportunity. *Local Government Studies* **5**, 57–90.

Flaherty, M. and Smit, B. 1982: An assessment of land classification techniques in planning for agricultural use. *Journal of Environment Management* **15**, 323–32.

Fliegel, F.C.; Sofranko, A.J. and Glasgow, N. 1981: Population growth in rural areas and sentiments of the new migrants towards further growth. *Rural Sociology* **46**, 411–29.

Foeken, D. 1980: Return migration to a marginal rural area in north western Ireland. *Tijdschrift voor Economische en Sociale Geografie* **71**, 114–20.

Forestry Commission 1967: *Census of woodlands 1965–67,* London: HMSO.

——1977: *The wood production outlook in Britain*. Edinburgh: The Commission.

——1983a: *62nd Annual Report and Accounts 1981–82. House of Commons Paper 363 (82–83),* London: HMSO.

——1983b: *Census of woodlands and trees 1979–82. England, Scotland, Wales.* 3 volumes. Edinburgh: The Commisson.

Forster, R.R. 1973: *Planning for man and nature in National Paks*. Geneva: IUCN.

Fothergill, S. and Gudgin, G. 1979: Regional employment change: A sub-regional explanation. *Progress in Planning* **12**, 155–219.

——1982: *Unequal growth: Urban and regional development change in the UK*. London: Heinemann.

Fotheringham, A.S. and Reeds, L.G. 1979: An application of discriminant analysis to agricultural land use prediction. *Economic Geography* **55**, 114-22.

Found, W.C. 1970: Towards a general theory relating distance between farm and home to agricultural production, *Geographical Analysis* **2**, 165-76.

——1971: *A theoretical approach to rural land use patterns*. London: Edward Arnold.

Franklin, S.H. 1971: *Rural societies*. London: Macmillan.

Fraser, A. 1982: The role of deciduous woodlands in the economy of rural communities. *Arboricultural Journal* **6**, 37-47.

Frost, M. and Spence, N. 1984: The changing structure and distribution of the British workforce. *Progress in Planning* **21**, 67-147.

Fuguitt, G.V. 1979: Population movements and integrated rural development. *Sociologia Ruralis* **19**, 99-115.

Gasson, J. 1969: The occupational immobility of small farmers. *Journal of Agricultural Economics* **20**, 279-88.

Gasson, R. 1973: Goals and values of farmers: *Journal of Agricultural Economics* **24**, 521-37.

——1974: Turnover and size of labour force on farms. *Journal of Agricultural Economics* **25**, 115-27.

——1975: *Provision of tied cottages*. Cambridge: Department of Land Economy.

Gayler, H. 1982: Conservation and development in urban growth: The preservation of agricultural land in the rural – urban fringe of Ontario. *Town Planning Review* **53**, 320-41.

Gibbs, R.S. 1976: *The impact of recreation on upland access land*. Newcastle-upon-Tyne: Agricultural Adjustment Unit.

Gilder, I.M. 1979: Rural planning policies: An economic appraisal. *Progress in Planning* **11**, pp.213-71.

Gilg, A.W. 1975a: Agricultural land classification in Britain: A review of the Ministry of Agriculture's new map series. *Biological Conservation* **7**, 73-7.

——1975b: Development control and agricultural land quality. *Town and Country Planning* **43**, 387-9.

——1976: Rural employment. In Cherry 1976, 125-72.

——1978: Policy forum: Needed: a new 'Scott' inquiry. *Town Planning Reivew* **49**, 353-71.

——1979: *Countryside planning. The first three decades 1945-76*. London: Methuen.

——1980a: Planning for rural employment in a changed economy. *Planner* **66**, 91-3.

——1980b: Planning for nature conservation: A struggle for survival and political respectability. In Kain, R. (ed.), *Planning for conservation* (London: Mansell), 97-116.

——1981: Provisional results of June 1980 Agricultural Census for UK *Countryside Planning Yearbook* **2**, 39.

——(ed.) 1982: *Countryside Planning Yearbook*. Norwich: GeoBooks.

——1983a: Provisional results of June 1982 Agricultural census for United Kingdom. *Countryside Planning Yearbook* **4**, 28-9.

——1983b: Belvoir development delayed, New Plans for Belvoir, Revised Belvoir plan accepted *Countryside Planning Yearbook* **4**, 20, 24 and 39.

——1983c: Population and employment: In Pacione 1983, 74-105.

——1984a: Plan to build 12 new villages. *Countryside Planning Yearbook* **5**, 25.

——1984b: No rural Minister. *Countryside Planning Yearbook* **5**, 32 and 109.

——1985. Rural Development Areas. *Countryside Planning Yearbook* **6**, forthcoming.

Gillon, S. 1981: Selling rural council houses. *Town and Country Planning* **50**, 115-16.

Glyn-Jones, A. 1977: *Village into town*. Exeter: Exeter University Press.

——1979. *Rural recovery: Has it begun*? Exeter: Exeter University Press.

——1982: *Small firms in a country town*. Exeter: Exeter University Press.

Goldschmidt, W. 1978: Large-scale farming and the rural social structure. *Rural Sociology* **43**, 362–6.

Goodall, B. 1973. The composition of forest landscapes. *Landscape Research News* **1**(5), 6–10.

Grafton, D. 1982: Net migration, outmigration and remote rural areas: A cautionary note. *Area* **14**, 313–18.

——1984: Small scale growth centres in remote rural regions: The case of Switzerland. *Applied Geography* **4**, 29–46.

Green, B. 1981: *Countryside conservation*. London: George Allen & Unwin.

Green, P.M. 1983: The impact of rural immigration on local government. In Brown and Hadwiger 1983, 83–97.

Gregor, H.F. 1970: *Geography of agriculture: Themes in research*. Englewood Cliffs: Prentice-Hall.

Gregory, D.G. 1970: *Green belts and development control*. Birmingham: Centre for Urban and Regional Studies.

Gregory, R. 1971: *The price of amenity*. London: Macmillan.

Grigg, D.B. 1974: *The agricultural systems of the world*. Cambridge: Cambridge University Press.

——1976: The world's agricultural labour force 1800–1970. *Geography* **361**, 194–202.

——1981: Agricultural geography. *Progress in Human Geography* **5**, 268–77.

——1982: Agricultural geography. *Progress in Human Geography* **6**, 242–6.

——1983: Agricultural Geography. *Progress in Human Geography* **7**, 255–60.

——1984: *An introduction to agricultural geography*. London: Hutchinson.

Grossman, D. 1971: Do we have a theory for settlement geography? *Professional Geographer* **23**, 197–203.

——1983: Commentary on 'Describing and Modelling Rural Settlement Maps' by Robert Haining. *AAAG* **73**, 298–300.

Guldmann, J.M. 1980: A mathematical experiment in landscape planning. *Environment and Planning B* **7**, 379–98.

Haffey, D. 1979. Recreational activity patterns on the uplands of an English National park. *Environmental Conservation* **6**, 237–42.

Hägerstrand, T. 1967: *Innovation diffusion as a spatial process*. Chicago: Chicago University Press.

Haining, R. 1982: Describing and modelling rural settlement maps. *AAAG* **72**, 211–23.

Hall, P. 1974: The containment of urban England. *Geographical Journal* **140**, 386–418.

Hall, P.; Thomas, R.; Gracey, H. and Drewett, R. 1973: *The containment of urban England*. London: George Allen & Unwin.

Halsall, D.A. and Turton, B.J. (eds.) 1979: *Rural transport problems in Britain: Papers and discussion*. Keele: IBG, Transport Geography Study Group.

Hamilton, P. and Woolcock, J. 1984: *Agricultural landscapes: An approach to their improvement: CCP 169*. Cheltenham: Countryside Commission.

Hannan, D.F. 1969: Migration motives and migration differentials among rural youth. *Sociologia Ruralis* **9**, 195–220.

Hanrahan, P.J. and Cloke, P.J. 1983: Towards a critical appraisal of rural settlement planning in England and Wales. *Sociologia Ruralis* **23**, 109–29.

Harkness, C.E. 1983: Mapping changes in the extent and nature of woodlands in National Parks in England and Wales. *Arboricultural Journal* **7**, 309–19.

Harman, R. 1978: Retailing in rural areas: A case study in Norfolk. *Geoforum* **9**, 107–26.

——1982: Rural services: Changes in north-east Norfolk. *Policy and Politics* **10**, 477–94.

Harris, R. 1979: Access to Landsat data. *Area* **11**, 63–6.

Harrison, C. 1981: A playground for whom? Informal recreation in London's Green Belt. *Area* **13**, 109–14.

——1983: Countryside recreation and London's urban fringe. *Transactions, IBG, New Series* **1**, 295–313.

Hart, J.F. 1976: Urban encroachment on rural areas. *Geographical Review* **66**, 1–17.
——1980: Land use change in a Piedmont county. *AAAG* **70**, 492–527.
Hart, P. 1977: Countryside implications of the decisions of rural landowners. *Area* **9**, 46–8.
Hart, P.W.E. 1978: Geographical aspects of contract farming with special reference to the supply of crops to processing plants. *Tijdschrift voor Economische en Sociale Geografie* **69**, 205–15.
——1980: Problems and potentialities of the behavourial approach to agricultural location. *Geografiska Annaler* **62B**, 99–108.
Harvey, D.W. 1963: Locational change in the Kentish hop industry and the analysis of land use patterns. *Transactions, IBG* **33**, 123–44.
——1966: Theoretical concepts and the analysis of agricultural land-use patterns in geography. *AAAG* **56**, 361–74.
——1969: *Explanation in geography.* London: Edward Arnold.
——1973: *Social justice and the city.* London: Edward Arnold.
Hassinger, E. 1957: The relationship of trade-center population change to distance from larger centers in an agricultural area. *Rural Sociology* **22**, 131–6.
Hathout, S.A. 1980: Mapping vegetation by Landsat image enhancement. *Journal of Environmental Management* **11**, 111–17.
Haynes, R.M. and Bentham, G.C. 1979: Accessibility and the use of hospitals in rural areas. *Area* **11**, 186–91.
HC 1978: *Innovations in rural bus services. House of Commons Paper 635 (77–78).* London: HMSO.
Healey, P. 1983: *Local plans in British land use planning.* Oxford: Pergamon.
Healy, R.G. and Rosenberg, J.S. 1980: *Land use and the States.* Baltimore: Johns Hopkins University Press.
Hedger, M. 1981: Re-assessment in Mid-Wales. *Town and Country Planning* **50**, 261–3.
HIDB 1983. *17th Annual Report.* London: HMSO.
Hirsch, G.P. and Maunder, A.H. 1978: *Farm amalgamation in Western Europe,* Farnborough: Saxon House.
HL Select Committee On Sport And Leisure 1973: *First and second reports.* London: HMSO.
HL Select Committee On The European Communities 1979: *Policies for rural areas in the European Community: House of Lords Paper 129 (79–80).* London: HMSO.
——1980. *The Common Agricultural Policy: House of Lords Paper 156 (79–80)* London: HMSO.
HL Select Committee On Science And Technology 1980. *Scientific Aspects of Forestry. House of Lords Paper 381 (79–80).* London: HMSO.
——1982. *Scientific aspects of forestry: Government response. House of Lords Paper 83 (81–82).* London: HMSO.
Hodge, I. and Whitby, M. 1981: *Rural Employment: Trends, options, choices.* London: Methuen.
——1982: The prospects and potential for rural employment. *Planning Outlook* **25**, 25–8.
Holtam, B. 1976: Forestry practice in Britain is applied terrestrial ecology. *Commonwealth Forestry Review* **55**, 123–7.
Hoskins, W.G. 1955: *The making of the English landscape.* London: Hodder & Stoughton. (Hoskins is also the editor of a regional set of texts based on counties published by Hodder & Stoughton.)
Housing and Local Government, Ministry of 1970: *Development plans: A manual on form and content.* London: HMSO.
Hudson, J.C. 1969: A location theory for rural settlement. *AAAG* **59**, 365–381.
Huggett, R. and Meyer, I. 1980: *Agriculture.* London: Harper & Row.
Hull, D. and Buyhoff, G. 1983: Distance and scenic beauty. *Environment and Behaviour* **15**, 77–91.

Ilbery, B.W. 1977: Point-score analysis: A methodological framework for analysing the decision making process in agriculture. *Tijdschrift voor Economische en Sociale Geografie* **68**, 66–71.
——1978: Agricultural decision-making: A behavioural perspective. *Progress in Human Geography* **2**, 448–66.
——1979: Decision making in agriculture: A case study of North-East Oxfordshire. *Regional Studies* **13**, 199–210.
——1981: Dorset agriculture: A classification of regional types. *Transactions, IBG, New Series* **6**, 214–27.
——1982: The decline of hop growing in Hereford and Worcester. *Area* **14**, 203–11.
——1983a: Harvey's principles re-applied: A case study of the declining West Midland hop industry. *Geoforum* **13**, 111–23.
——1983b: Goals and values of hop farmers. *Transactions, IBG*, New Series, **8**, 329–41.
——1983c: A behavioural analysis of hop farming in Hereford and Worcestershire. *Geoforum* **14**, 447–60.
Ilbery, B.W. and Hornby, R. 1983: Repertory grids and agricultural decision making: A mid Warwickshire case study. *Geografiska Annaler* **65B**, 77–84.
Irving, B. and Hilgendorf, L. 1975: *Tied Cottages in British Agriculture*. London: Tavistock Institute.
ITE 1978: *Upland land use in England and Wales*. Cheltenham: Countryside Commission
IUCN 1980. *World Conservation Strategy*. Geneva: IUCN.
Jackson, R.H. 1980: *Land Use in America*. London: Edward Arnold.
Jacques, D.L. 1980: Landscape appraisal: The case for a subjective theory. *Journal of Environmental Management* **10**, 107–13.
Johansen, H.E. and Fuguitt, G.U. 1984: *The changing rural village in America*. Cambridge, Mass. Ballinger.
Johnson, H.B. and Pitzl, G.R. 1981: Viewing and perceiving the rural scene.*Progress in Human Geography* **5**, 211–33.
Johnston, R.J. 1965: Components of rural population change. *Town Planning Review* **36**, 279–93.
——1971: Resistance to migration and the mover/stayer dichotomy: Aspects of kinship and population stability in an English rural area. *Geografiska Annaler* **53B**, 16–27.
——1979: *Geography and geographers*, London: Edward Arnold.
Johnston, R.J. and Doornkamp, J. (eds.) 1983: *The changing geography of the United Kingdom*. London: Methuen.
Jones, A. 1975: *Rural housing: The agricultural tied cottage*. London: Bell.
Jones, D.W. 1982: Location and land tenure. *AAAG* **72**, 314–31.
——1983: A neo-classical land use model with production, consumption and exchange. *Geographical Analysis* **15**, 128–41.
——1984a: Location, agricultural risk and farm income diversification. *Geographical analysis* **15**, 231–46.
——1984b: Farm location and off farm employment. *Transactions, IBG* **9**, 106–23.
——1984c: A land use model with a constant utility spatial variant wage. *Geographical Analysis* **16**, 121–33.
Jones, H.R. 1965: A study of rural migration in Central Wales. *Transactions, IBG* **37**, 31–45.
——1976: The structure of the migration process: Findings from a growth point in mid-Wales. *Transactions, IBG* New Series **1**, 421–32.
——1981: *A Population Geography*. London: Harper and Row.
Jones H.R.; Caird, J.B.; Berry, W.G. and Ford, N.J. 1984: *Counter-urbanisation: English migration to the Scottish Highlands and Islands*. In Jones, H. (ed.) *Population change in contemporary Scotland* (Norwich: GeoBooks), 71–83.
Jones, P. 1980: Primary school provision in rural areas. *The Planner* **66**, 4–6.
Joseph, A.E. and Smit, B. 1983: Preferences for public service provision in rural areas

undergoing ex-urban residential development: A Canadian view. *Tijdschrift voor Economische en Sociale Geografie* **74**, 41–52.

Kariel, H.G. and Kariel, P.E. 1982: Socio-cultural impacts of tourism: An example from the Austrian Alps. *Geografiska Annaler* **64B**, 1–16.

Keeble, D.; Owens, P.L. and Thompson, C. 1983: The urban – rural manufacturing shift in the European Community. *Urban Studies* **20**, 405–18.

Kennett, S. and Spence, N. 1979: British population trends in the 1970s. *Town and Country Planning* **48**, 220–23.

Killeen, K. and Buyhoff, G. 1983: The relation of landscape preferences to abstract topography. *Journal of Environmental Management* **17**, 381–92.

Knowles, R. 1978: Requiem for the rural bus. *Geographical Magazine* **50**, 668–71.

Knox, P.L. and Cottam, M.B. 1981a: Rural deprivation in Scotland: A preliminary assessment. *Tjidschrift voor Economische en Sociale Geografie* **72**, 162–75.

——1981b: A welfare approach to rural geography: Contrasting perspectives on the quality of Highland life. *Transactions, IBG New Series* **6**, 433–50.

Kostrowicki, J. 1979: Twelve years activity of the IGU Commission on agricultural typology. *Geographica Polonica* **40**, 235–60.

——1982: Types of agriculture map of Europe. *Geographica Polonica* **48**, 79–92.

Kozlowski, J. and Hughes, J.T. 1972: *Threshold analysis*. London: Architectural Press.

Kreimer, A. 1977: Environmental preferences: A critical analysis of some research methodologies. *Journal of Leisure Research* **9**, 88–97.

Kusler, J.A. 1980: *Regulating sensitive lands*. Cambridge, Mass.: Ballinger.

Labour Party 1981: *Out of town, out of mind: A programme for rural revival*. London: Labour Party.

Lang, J.T. 1982: The European Community directive on bird conservation. *Biological Conservation* **22**, 11–25.

Lapping, M.B. 1977: Policy alternatives for the preservation of agricultural land use. *Journal of Environmental Management* **5**, 275–87.

Lapping, M.B. and Clemenson, H.A. 1984: Recent developments in North American rural planning. *Countryside Planning Yearbook* **5**, 42–61.

Lassey, W.R. 1977: *Planning in rural environments*. New York: McGraw-Hill.

Lavery, P. (ed.) 1974: *Recreational Geography*. Newton Abbot: David & Charles.

——1975: The demand for recreation. *Town Planning Review* **46**, 185–200.

Law, C.M. and Warnes, A.M. 1976: The changing geography of the elderly in England and Wales. *Transactions, IBG, New Series* **1**, 453–71.

Lawton, R. 1968: The journey to work in Britain: Some trends and problems. *Regional Studies* **2**, 23–40.

Leat, D. 1981: The role of pressure groups in rural planning. *Countryside Planning Yearbook* **2**, 71–83.

Lefaver, S. 1978: A new framework for rural planning. *Urban Land* **37**, 7–13.

Leonard, P. 1980: Agriculture in the National Parks of England and Wales: A conservation viewpoint. *Landscape Planning* **7**, 369–86.

——1982: Management agreements: A tool for conservation. *Journal of Agricultural Economics* **33**, 351–60.

Leonard, P. and Cobham, R.O. 1977: The farming landscapes of England and Wales: A changing scene. *Landscape Planning* **4**, 205–36.

Lewis, G.J. 1967: Commuting and the village in mid-Wales. *Geography* **52**, 294–304.

——1979: *Rural communities*. Newton Abbot: David & Charles.

——1982: *Human migration*. London: Croom Helm.

——1983. Rural communities. In Pacione 1983, 149–72.

Lewis G.J. and Maund, D.J. 1976: The urbanisation of the countryside: A framework for analysis. *Geografiska Annaler* **58B**, 17–27.

——1979: Intra-community segregation: A case study in rural Herefordshire. *Sociologia Ruralis* **19**, 135–47.

Liberal Party 1983: *Liberal country: A manifesto for rural Britain.* London: Liberal Party.
Liddle, M.J. and Scorgie, H.R.A. 1980: The effects of recreation on freshwater plants and animals. A review. *Biological Conservation* 17, 183–206.
Linton, D.L. 1968: The assessment of scenery as a natural resource. *Scottish Geographical Magazine* 84, 219–38.
de Lisle, D.G. 1982: Effects of distance on cropping patterns internal to the farm. *AAAG* 72, 88–98.
Lloyd, R. 1979: Agricultural landscapes. *Landscape Research* 4, 18–22.
Lonsdale, R. 1981: Drawing conclusions from an examination of two nations. In Lonsdale and Holmes 1981, 337–86.
Lonsdale, R. and Browning, C. 1971: Rural – urban locational preferences of southern manufacturers. *AAAG* 61, 255–68.
Lonsdale, R. and Holmes, J. (eds.) 1981: *Settlement systems in sparsely populated regions.* Oxford: Pergamon.
Losch, A. 1954: *The economics of location.* Translated from the 1943 publication 'Die Räumliche Ordnung der Wirtschaft' by W.H. Woglom, New Haven: Yale University Press.
Lovejoy, D. 1983: Trees, landscape and conservation, *Landscape Research* 8, 25–9.
Low, N. 1973: Farming and the inner green belt. *Town Planning Review* 44, 103–15.
Lowe, P. and Goyder, J. 1983: *Environmental groups in British politics.* London: George Allen & Unwin.
Lowenthal, D. 1978: Finding valued landscapes. *Progress in Human Geography* 2, 373–418.
Lyons, E. 1983: Demographic correlates of landscape preference. *Environment and Behaviour* 15, 487–511.
McCallum, J.D. and Adams, J.G.L. 1981: Employment and unemployment statistics for rural areas. *Town Planning Review* 52, 157–66.
McCoy, E.D. 1983: The application of island-biogeographic theory to patches of habitat: How much land is enough? *Biological Conservation* 25, 53–61.
McCuaig, J. and Manning, E. 1982: *Agricultural land use change in Canada.* Ottawa: Lands Directorate.
MacEwen, M. and MacEwen, M. 1981: *National Parks: Conservation or cosmetics.* London: George Allen & Unwin.
——1982: The Wildlife and Countryside Act 1981: An unprincipled Act? *The Planner* 68, 69–71.
McLaughlin, B.P. 1976: Rural settlement planning: A new approach. *Town and Country Planning* 44, 156–60.
——1981: Rural deprivation. *The Planner* 67, 31–3.
McLaughlin, B.P. and Singleton, P. 1979: Recreational use of a nature reserve: A case study in North Norfolk, U K *Journal of Environmental Management* 9, 213–23.
MacLeary, A.R. 1981: Rural planning: Problems and policies. *Journal of Agricultural Economics* 32, 317–29.
McNab, A. 1984: *Integrated rural development.* Gloucester: Gloucestershire College of Arts.
McNamara, P. and Healey, P. 1984: The limitations of development control data in planning research: A comment on Ian Brotherton's recent study. *Town Planning Review* 55, 91–101.
Mabogunje, A. 1970: Systems approach to a theory of rural – urban migration. *Geographical Analysis* 2, 1–18.
MAFF 1968: *A century of agricultural statistics 1866–1966,* London: HMSO.
——1974: *Type of farm maps of England and Wales: Explanatory note,* London: HMSO, *Maps,* Alnwick, MAFF.
——1976: *Wildlife conservation in semi-natural habitats on farms,* London: HMSO.
——1977: *The changing structure of agriculture 1968–75,* London: HMSO.

——1980: *Farm classification in England and Wales 1976–77*, London: HMSO.

——1983a: *Annual review of agriculture: Cmnd 8804*, London: HMSO.

——1983b: *Agricultural statistics United Kingdom 1980–81*, London: HMSO.

Mallet, A. 1978: Agriculture et tourisme dans un milieu haut-alpin: Un example brianconnais. *Etudes Rurale* **71–72**, 111–54.

Mandelker, D.R. and Cunningham, R.A. 1979: *Planning and control of land development*. Indianapolis: Bobbs-Merrill.

Manners, G. 1981: *Coal in Britain: An uncertain future*. London: George Allen & Unwin.

Maos, J. 1983: The efficiency of services in dispersed and concentrated land settlement: A comparison. *Geografiska Annaler* **65B**, 47–56.

Margules, C,; Higgs, A.J. and Rafe, R.W. 1982: Modern biogeographic theory: Are there any lessons for nature reserve design? *Biological Conservation* **24**, 115–28.

Marshall, I.B. 1982: *Mining, land use and the environment*. Ottawa: Lands Directorate Environment Canada.

Martin and Voorhees Associates 1981: *Review of rural settlement policies 1945–80*. Bristol: Department of the Environment.

Martin, W.H. and Mason, S. 1976: Leisure 1980 and beyond. *Long Range Planning* **9**, 58–65.

Mather, A.S. 1978: Patterns of afforestation in Britain since 1945: *Geography* **63**, 157–66.

Matthieu, N. and Bontron, J.C. 1973: Les transformations de l'espace rurale. *Etudes Rurale* **49–50**, 137–59.

Maxwell, T.J.: Sibbald, A.R. and Eadie, J. 1979: Integration of forestry and agriculture: A model. *Agricultural Systems* **4**, 161–88.

Meinig, D.W. (ed.) 1979: *The interpretation of ordinary landscapes*. Oxford: Oxford University Press.

Merrett, S. 1984: Villages which have an appetite for land. *Town and Country Planning* **53**, 140–2.

Miles, J.C. and Smith, N. 1977: *Models of recreational traffic in rural areas*. Crowthorne: Transport and Road Research Laboratory.

Miles, R. 1967: *Forestry in the English landscape*. London: Faber.

Miller, M.K. and Luloff, A.E. 1981: Who is rural? A typological approach to the examination of rurality. *Rural Sociology* **46**, 608–25.

Minay, C. 1977: A new rural development agency. *Town and Country Planning* **45**, 439–43.

Minister Of Land And Natural Resources 1966: *Leisure in the countryside: Cmnd 2928*. London: HMSO.

Mitchell, G.D. 1950: Depopulation and the rural social structure. *Sociological Review* **42**, 69–85.

——1951: The relevance of group dynamics to rural planning problems. *Sociological Review* **43**, 1–16.

Mogey, J. 1976: Recent changes in the rural communities of the United States. *Sociologia Ruralis* **16**, 139–60.

M O H. 1944: *Rural housing*. London: HMSO.

——1948: *Housing survey in rural areas*. London: HMSO.

Moran, W. 1979: Spatial patterns of agriculture on the urban periphery. The Auckland Case. *Tijdschrift voor Economische en Sociale Geografie* **70**, 164–76.

Morgan, W.B. and Munton, R.J.C. 1971: *Agricultural geography*. London: Methuen.

Moseley, M.J. 1973a: Some problems of small expanding towns. *Town Planning Review* **44**, 263–78.

——1973b: The impact of growth centres in rural regions – I. An analysis of spatial 'patterns' in Brittany. *Regional Studies* **7**, 57–75.

——1973c: The impact of growth centres in rural regions – II. An analysis of spatial 'flows' in East Anglia. *Regional Studies* **7**, 77–94.

——(ed.) 1978: *Social issues in rural Norfolk.* Norwich: Centre for East Anglian Studies.

——1979: *Accessibility: The rural challenge.* London: Methuen.

——1980: Is deprivation really rural? *The Planner* **66**, 97.

——1983. The rural areas. In Johnston and Doornkamp 1983, 257–74.

Moseley, M.J. and Darby, J. 1978: The determinants of female activity rates in rural areas. An analysis of Norfolk parishes. *Regional Studies* **12**, 297–309.

Moseley, M.J.; Harman, R.G.; Coles, O.B. and Spencer, M.B. 1977: *Rural transport and accessibility.* Norwich: Centre for East Anglian Studies.

Moss, G. 1978: The village. A matter of life or death. *Architects Journal* **167**, 100–39.

——1980: *Trent Study: Planning for change in a rural community.* Fairford: Ernest Cook Trust.

Muir, R. 1981: *The shell guide to reading the landscape.* London: Michael Joseph.

Muller, P.O. 1973: Trend surfaces of American agricultural patterns: A macro-Thünian analysis. *Economic Geography* **49**, 228–42.

Munton, R. 1982: Green Belt Policy: What role for agriculture? *Planning Outlook* **25**, 43–51.

——1983: *London's Green Belt: Containment in practice.* London: George Allen & Unwin.

Munton, R. and Norris, J.M. 1969: The analysis of farm organisation: An approach to the classification of agricultural land in Britain. *Geografiska Annaler* **53B**, 95–103.

National Consumer Council 1978: *Rural rides: Experiments in rural pulic transport: A consumer view.* London: The Council.

NCC 1984: *Nature conservation in Great Britain.* Shrewsbury: NCC.

NCVO 1980a: *Jobs in the countryside.* London: NCVO.

——1980b: *Rural futures: Impact of new technology on the countryside.* London: NCVO.

Neate, S. 1981: *Rural deprivation: An annotated bibliography.* Norwich: GeoBooks.

Nelson, J.G. (ed.) 1978: *International experience with national parks and related reserves.* Waterloo: Department of Geography.

Netting R.M. 1981: *Balancing on an alp: Ecological change and continuity in a Swiss mountain community.* Cambridge: Cambridge University Press.

de Neufville, J.I. (ed.) 1981: *The land use policy debate in the United States.* New York: Plenum Press.

Newbury, P.A.R. 1980: *A geography of agriculture.* Plymouth: Macdonald & Evans.

Newby, H. 1977: *The deferential worker.* London: Allen Lane.

——1979: *Green and pleasant land: Social change in rural England.* London: Pelican.

Newby, H.; Bell, C.; Rose, D. and Saunders, P. 1978: *Property paternalism and power: Class and control in rural England.* London: Hutchinson.

Newby, H.; Bell; C.; Saunders, P. and Rose, P. 1977. Farmers' attitudes to conservation. *Countryside Recreation Review* **2**, 23–30.

Nicholls, D.C. 1969: *Use of land for forestry within the proprietary land unit.* London: HMSO.

Nicholson, B. 1975: Return migration to a marginal rural area: An example from north Norway. *Sociological Ruralis* **15**, 227–44.

Norfolk County Council 1976: *North Walsham area: A case study of alternative patterns of development.* Norwich: Norfolk County Council.

Norfolk County Planning Department 1979: *Rural areas in England and Wales.* Norwich: Norfolk County Council.

North, J. and Spooner, D.J. 1978: On the coalmining frontier. *Town and Country Planning* **46**, 155–63.

——1982: A future for coal. *Town and Country Planning* **51**, 93–8.

Nutley, S.D. 1979: Patterns of regional accessibility in the N.W. Highlands and Islands. *Scottish Geographical Magazine* **95**, 142–54.

——1980a: Accessibility, mobility and transport related welfare: The case of rural Wales. *Geoforum* **11**, 335–52.

——1980b: The concept of 'isolation' – A method of evaluation and a West Highland example. *Regional Studies* **14**, 111–23.

OECD 1976: *Land use policies and agriculture*. Paris: OECD.

——1977: *Review of agricultural policies*. Paris: OECD.

——1983: *Review of agricultural policies in OECD member countries*. Paris: OECD.

Olwig, K. 1984: *Nature's ideological landscape*. London: George Allen & Unwin.

OPCS 1980: *Population density and concentration in Great Britain, 1951, 1961 and 1971*. London: HMSO.

——1981a: *Census 1981: Preliminary report*, London: HMSO.

——1981b: The first results of the 1981 Census of England and Wales. *Population Trends* **25**, 1–9.

——1984: In 1984 the office published: Annual Estimates under the series *PP1*; Quarterly Estimates in the Journal *Population Trends*; and Monthly estimates of births and deaths to individual subscribers.

Openshaw, S.; Cullingford, D. and Gillard, A. 1980: A critique of the national classifications of OPCS/PRAG. *Town Planning Review* **51**, 421–39.

O'Riordan, T. 1982: *Putting trust in the countryside: Earth's survival: A conservation and development programme for the U K* London: NCC.

Orwin, C.S. 1944: *Country planning*. Oxford: Oxford University Press.

Owen, D.W.; Coombes, M.G. and Gillespie, A.E. 1983: *The differential performance of urban and rural areas in the recession*. Newcastle upon Tyne: Centre for Urban and Regional Development Studies.

Owens, P.L. 1984: Rural leisure and recreation research: A retrospective evaluation. *Progress in Human Geography* **8**, 157–88.

Pacione, M. 1980: Quality of life in a metropolitan village. *Transactions, IBG, New Series* **5**, 185–206.

——1982: The viability of smaller rural settlements. *Tijdschrift voor Economische en Sociale Geografie* **73**, 149–61.

——(ed.) 1983: *Progress in rural geography*, London: Croom Helm.

——1984: *Rural geography*. London: Harper & Row.

Pahl, R.E. 1965: *Urbs in Rure*. London School of Economics.

——1966a: The social objectives of village planning. *Official Architecture and Planning* **29**, 1146–1150.

——1966b: The rural – urban continuum. *Sociologia Ruralis* **6**, 299–329.

Palmer, C.J.; Robinson, M.E. and Thomas, R.W. 1977: The countryside image: An investigation of structure and meaning. *Environment and Planning A* **9**, 739–49.

Parry, M.L. 1976: The mapping of abandoned farmland in upland Britain. *Geographical Journal* **142**, 101–10.

——1983: The changing use of land. In Johnston and Doornkamp 1983, 13–36.

Parry, M.L.; Bruce, A. and Harkness, C.E. 1982: *Surveys of moorland and roughland change*. Birmingham: Department of Geography, University of Birmingham.

Patmore, J.A. 1970: *Land and leisure*. Newton Abbot: David & Charles.

——1983: *Recreation and resources: Leisure patterns and leisure places*. Oxford: Basil Blackwell.

Peak Park Joint Planning Board 1976: *Structure Plan*. Bakewell: The Board.

——1978: *National Park Plan*. Bakewell: The Board.

——1984: *A tale of two villages*. Bakewell: The Board.

Peake, H. 1918: The regrouping of rural population. *Town Planning Review* **7**, 243–50.

Peat, Marwick, Mitchell & Co. 1980: *The assessment of needs for rural public transport* London: Peat, Marwick, Mitchell.

Penning-Rowsell, E.C. 1974: Landscape evaluation for development plans. *The Planner* **60**, 930–4.

——1975: Constraints on the application of landscape evaluations. *Transactions, IBG* **66**, 149–55.

——1981: Fluctuating fortunes in gauging landscape value. *Progress in Human Geography* **5**, 25–41.

——1982: A public preference evaluation of landscape quality. *Regional Studies* **16**, 97–112.

——1983: County landscape conservation policies in England and Wales. *Journal of Environmental Management* **16**, 211–28.

Penoyre, J. and Penoyre, J. 1978: *Houses in the landscape*. Newton Abbot: Readers Union.

Peterken, G.F. 1981: *Woodland conservation and management*. London: Chapman & Hall.

Peterken, G.F. and Harding, P.T. 1975: Woodland conservation in Eastern England: Comparing the effects of changes in three study areas since 1946. *Biological Conservation* **8**, 279–98.

Peters, G.H. 1970: Land use studies in Britain. *Journal of Agricultural Economics* **21**, 171–213.

Peterson, G. 1983: Methods for retaining agriculture land in the urban fringe in the USA. *Landscape Planning* **9**, 271–8.

Phillips, D. and Williams, A. 1982a: Local authority housing and accessibility: evidence from the South Hams, Devon. *Transactions, IBG New Series* **7**, 304–20.

——1982b: *Rural housing and the public sector*. Aldershot: Gower.

——1983a: Rural settlement policies and local authority housing. *Environment and Planning A* **15**, 501–13.

——1983b: The social implications of rural housing policy. *Countryside Planning Yearbook* **4**, 77–102.

——1984: *Rural Britain: A social geography*. Oxford: Basil Blackwell.

Picou, J.S.; Weils, R.H. and Nyberg, K.L. 1978: Paradigms, theories and methods in contemporary rural sociology. *Rural Sociology* **43**, 559–83.

Pierce, J.T. 1981: Conversion of rural land to urban: A Canadian profile. *Professional Geographer* **32**, 163–73.

Pigram, J. 1983: *Outdoor recreation and resource management*. Beckenham: Croom Helm.

Platt, R.H. 1981: Farmland conversion: National lessons for Iowa. *Professional Geographer* **33**, 113–21.

Pocock, D.C.D. 1960: Regional patterns of English hop growing. *Tijdschrift voor Economische en Sociale Geografie* **51**, 108–14.

——(ed.) 1981: *Humanistic geography and literature*. London: Croom Helm.

Porchester, Lord, 1977: *A study of Exmoor*. London: HMSO.

Potter, C. 1983: *Investing in rural harmony: An alternative package of agricultural subsidies and incentives for England and Wales*. Godalming: World Wildlife Fund.

Preece, R.A. 1980: *An evaluation by the general public of scenic quality in the Cotswolds Area of Outstanding Natural Beauty*. Oxford: Oxford Polytechnic.

Price, C. 1976: Subjectivity and objectivity in landscape evaluation. *Environment and Planning A* **8**, 829–38.

——1979: Public preference and the management of recreational congestion. *Regional Studies* **13**, 125–40.

Price, C. and Dale, I. 1982: Price predictions and economically afforestable area. *Journal of Agricultural Economics* **33**, 13–23.

Probert, G. and Hamersley, C. 1979: Countryside management in Gwent. *The Planner* **65**, 10–13.

Propst, D.B. and Buyhoff, G.J. 1980: Policy capturing and landscape preference quantification. *Journal of Environmental Management* **11**, 45–59.

Rackham, O. 1976: *Trees and Woodland in the British landscape*. London: J.M. Dent.

Radford, E. 1970: *The new villagers*. London: Cass.

Ramblers Association 1980: *Afforestation: The case against expansion*. London: The Association.

Randolph, W.G. and Robert, S. 1981: Population redistribution in Great Britain 1971–1981. *Town and Country Planning* **50**, 227–30.

Ravenstein, E.G. 1885: The laws of migration. *Journal of the Royal Statistical Society* **48**, 167–235.

Rayner, A.J. 1977: The regional pricing policy of the Milk Marketing Board and the public interest. *Journal of Agricultural Economics* **28**, 11–25.

Rees, G. and Rees, T.L. 1977: Alternatives to the Census: The example of sources of internal migration data. *Town Planning Review* **48**, 123–40.

Relph, E. 1981: *Rational landscapes and humanistic geography*. London: Croom Helm.

Rettig, S. 1976: An investigation into the problems of urban fringe agriculture in a green belt situation. *Planning Outlook,* **19**, 50–74.

Rhind, D. (ed.) 1983: *A census users handbook*. London: Methuen.

Rhind, D. and Hudson, R. 1980: *Land Use*. London: Methuen.

Richmond, P. 1983: *Housing associations in rural areas*. Plymouth: South West Papers in Geography.

Rieger, J.H. 1972: Geographic mobility and the occupational attainment of rural youth: A longitudinal study. *Rural Sociology* **37**, 189–207.

Robert, S. and Randolph, W.G. 1983: Beyond decentralisation: The evolution of population distribution in England and Wales 1961–81. *Geoforum* **14**, 75–102.

Roberts, B.K. 1977: *Rural settlement in Britain*. Folkestone: Dawson.

Robertson, I.M.L. 1961: The occupational structure and distribution of rural population in England and Wales. *Scottish Geographical Magazine* **77**, 165–79.

——1976: Accessibility to services in the Argyll district of Strathclyde: A locational model. *Regional Studies* **10**, 89–95.

Robinson, D.G.; Wager, J.F.; Laurie, I.C. and Traill, A.L. 1976: *Landscape evaluation*. Manchester: University of Manchester.

Rogers, A. 1981a: Housing in the national parks. *Town and Country Planning* **50**, 193–5.

——(ed.) 1981b: Landscape evaluation in planning rural areas. *Landscape Research* **6**, 1–25.

——1983: Housing. In Pacione 1983, 106–29.

Rogers, E.M. and Burdge, R.J. 1972: *Social change in rural societies*. Englewood Cliffs: Prentice-Hall.

Roome, N.J. 1983: Preferences of national nature reserve users. *Journal of Environmental Management* **17**, 143–52.

Rowley, G. 1971: Central places in rural Wales. *AAAG* **61**, 537–50.

Rowley, T. 1978: *Villages in the landscape*. London: Dent.

Royal Geographical Society 1983: *Minutes of evidence submitted to the Select Committee on Science and Technology: Remote Sensing and Digital Mapping: House of Lords Paper 127-iii (82–83)*. London: HMSO.

Runte, A. 1979: *National parks: The American experience*. Lincoln: University of Nebraska Press.

Rural Resettlement Group 1979: *Rural resettlement handbook*. Diss: The Group.

Sargent, F.O. 1976: *Rural environment planning*. Vermont: University of Vermont.

Saville, J. 1957: *Rural depopulation in England and Wales 1851–1951*. London: Routledge & Kegan Paul.

Schaefer, F.K. 1953: Exceptionalism in geography. *AAAG* **43**, 226–49.

SDD 1972: *Threshold Analysis Manual*. London: HMSO.

Secretary Of State For The Environment 1975: *Sport and recreation: Cmnd 6200* London: HMSO.

Selman, P. 1977: Approaches to the multiple use of the uplands. *Town Planning Review* **49**, 163–74.

——1982: The use of ecological evaluations by local planning authorities. *Journal of Environmental Management* 10, 139–47.

Settle, J.G. 1980: Relating participation in recreational activities to social characteristics, *Journal of Environmental Management* 10, 139–47.

Shaw, D. 1984: A tide thats turning in the rural west. *Town and Country Planning* 52, 255–6.

Shaw, J.M. 1976: Can we afford villages? *Built Environment* 2, 135–7.

Shaw, M.J. (ed.) 1979: *Rural deprivation and planning*. Norwich: GeoBooks.

Shoard, M. 1980: *The theft of the countryside*. London: Temple Smith.

Shucksmith, D.M. 1980a: Local interests in a national park. *Town and Country Planning* 49, 418–21.

——1980b: Petrol prices and rural recreation in the 1980s. *National Westminster Bank Quarterly Review* February, 52–9.

——1981: *No homes for locals?* Aldershot: Gower.

——1983: Second homes: A framework for policy. *Town Planning Review* 54, 174–93.

Shuttleworth, S. 1980a: The evaluation of landscape quality. *Landscape Research* 5, 14–15, 18–29.

——1980b: The use of photographs as an environment presentation medium in landscape studies. *Journal of Environmental Management* 11, 61–76.

Simmons, I. 1975: *Rural recreation in the industrial world*. London: Edward Arnold.

——1980: Ecological-functional approaches to agriculture in geographical contexts. *Geography* 65, 305–6.

Simpson, B. 1977: Some theoretical developments, a threshold analysis and testing. *Urban Studies* 14, 79–88.

Simpson, E.S. 1959: Milk production in England and Wales. *Geographical Review* 49, 95–111.

Sinclair, G. (ed.) 1983: *The Upland Landscapes Study*. Narbeth: Environment Information Services.

Sinden, J.A. 1976: Carrying capacity as a planning concept for national parks: Available or desirable capacity. *Landscape Planning* 2, 243–7.

Slee, R.W. 1981: Agricultural policy and remote rural areas. *Journal of Agricultural Economics* 32, 113–21.

——1982: *County Parks: A review of policy and management issues*. Gloucester: Gloucestershire College of Arts.

Sluman, B. 1984: Can conservation be a part of farming? *Europe* 9, 12.

Smart, G. and Wright, S., 1983: *Decision making for rural areas*. London: Bartlett School of Architecture and Planning.

Smith, S.L.J. 1983: *Recreation geography*. London: Longmans.

South West Economic Planning Council 1975: *Survey of second homes in the South West*. London: HMSO.

——1976: *Economic survey of the tourist industry in the South West*. London: HMSO.

Spooner, D.J. 1972: Industrial movement and rural periphery: The case of Devon and Cornwall. *Regional Studies* 6, 197–215.

——1981a: *Mining and regional development*. Oxford: Oxford University Press.

——1981b: The geography of coal's second coming. *Geography* 66, 29–41.

Stamp, L.D. 1962: *The land of Britain: It's use and misuse*, 3rd ed. London: Longmans.

Standing Conference Of RCCs 1978: *The decline of rural services*. London: The Conference.

Stanley, P.A. and Farrington, J.H. 1981: The need for rural public transport. A constraints based study. *Tijdschrift voor Economische en Sociale Geografie* 72, 62–80.

Stewart, A.J.A. and Lance, A.N. 1983: Moor-draining: A review of impacts on land use. *Journal of Environmental Managment* 17, 81–99.

Strauss, E. and Churcher, E.H. 1967: The regional analysis of the milk market. *Journal of Agricultural Economics* 18, 221–40.

Strong, A.L. 1983: Easements as a development control in the United States. *Landscape Planning* **10**, 43–64.

Sturrock, F.G. and Cathie, J. 1980: *Farm modernisation and the countryside.* Cambridge: Department of Land Economy.

Symons, L. 1978: *Agricultural geography.* London: Bell & Hyman.

Tandy, C. 1978: Forestry and recreation. *Landscape Design* **124**, 11–12.

Tarrant, J.R. 1974: *Agricultural geography.* Newton Abbot: David & Charles.

——1980a: Agricultural trade within the European Community. *Area* **12**, 37–42.

——1980b: Production and disposal of surplus EEC milk products. *Area* **12**, 247–52.

Tavener, L.E. 1952: Changes in the agricultural geography of Dorset. *Transactions, IBG* **18**, 93–106.

Taylor, C. and Emerson, D. 1981: *Rural post offices: Retaining a vital service.* London: Bedford Square Press.

Thissen, F. 1978: Second homes in the Netherlands. *Tijdschrift voor Economische en Sociale Geografie* **69**, 323–32.

Thomson, K.J. 1981: *Farming in the urban fringe*, Cheltenham: Countryside Commission.

Thorns, D.C.1968: The changing system of social stratification. *Sociologia Ruralis* **8**, 161–77.

—— 1970: Participation in rural planning. *International Review of the Community Development* **24**, 129–38.

Thurgood, G. 1978: Rural housing initiatives: Are we doing enough? *The Planner* **64**, 143–5.

Tonnies, F. 1887: *Gemeinschaft und Gesellschaft.* Translated and supplemented by C. Loomis (1957) Community and Society. Ann Arbor: Michigan State University Press.

Towler, R.W. 1975: Forestry in national parks. *Quarterly Journal of Forestry* **69**, 129–36.

Tracy, M.A. 1976: Fifty years of agricultural policy. *Journal of Agricultural Economics* **27**, 331–48.

Transport, Department of 1984: *Buses.* London: HMSO.

Treasury, H.M. 1972: *Forestry policy.* London: HMSO.

——1976: *Rural depopulation.* London: The Treasury.

Trimble, S.W. 1983: Commentary on 'Land Use Change in a Piedmont County' by John Fraser Hart. *AAAG* **73**, 285–7.

TRRU 1975 to 1977. *The Scottish tourism and recreation study. 1: Survey description; 2: Summary report; 3: Holidaymaking in Scotland; 4: The woodland visitor.* Edinburgh: TRRU.

——1981: *The economy of rural communities in the National Parks of England and Wales.* Edinburgh: TRRU.

——1983: *Recreation site survey manual: Methods and techniques for conducting visitor surveys.* London: Spon

Turnock, D. 1968: Depopulation in north-east Scotland with reference to the countryside. *Scottish Geographical Magazine* **84**, 256–68.

Usher, M.B. and Miller, A.K. 1975: The development of a nature reserve as an area of conservational and recreational interest. *Environmental Conservation* **2**, 202–4.

Vince, S.W.E. 1952: Reflections on the structure and distribution of rural population in England and Wales 1921–31. *Transactions, IBG* **18**, 53–76.

Vining, D. and Pallone, R. 1982: Migrations between core and peripheral regions: A description and tentative explanation of the pattterns in 22 countries. *Geoforum* **13**, 339–410.

Visser, S. 1980: Technological change and the spatial structure of agriculture. *Economic Geography* **56**, 311–19.

——1982: On agricultural location theory. *Geographical Analysis* **14**, 167–76.

Von Thünen, J.H. 1966: *Von Thünen's Isolated State* edited by P. Hall from: Der isolierte Staat in Beziehung auf Landwirtschaft und National ëkonomie. London: Pergamon.

Wagstaff, H.R. 1974: The mobility replacement and wage rates of farm workers. *Oxford Agrarian Studies* **III**, 140–53.

Walker, A. (ed.) 1978: *Rural poverty: Poverty deprivation and planning in rural areas.* London: Child Poverty Action Group.

Wardwell, J.M. 1977: Equilibrium and change in non-metropolitan growth. *Rural Sociology* **42**, 156–79.

Warford, J. 1969: *The South Atcham Scheme: An economic appraisal.* London: HMSO.

Warner, W.K. 1974: Rural society in a post-industrial age. *Rural Sociology* **39**, 306–18.

Watkins, C. and Wheeler, P.T. (eds.) 1981: *The study and use of British woodlands.* Nottingham: Rural Geography Study Group.

Weaver, J.C. 1954: Cop combination regions in the middle west. *Geographical Review* **44**, 175–200.

Webber, R.J. 1980: A response to the critique of the national classification of OPCS/PRAG. *Town Planning Review* **51**, 440–50.

Webber, R.J. and Craig, J. 1978: *Socio-economic classification of local authority areas.* London: HMSO.

Wellman, J.D. and Buyhoff, G.J. 1980: Effects of regional familiarity on landscape preferences. *Journal of Environmental Management* **11**, 105–10.

Westmacott, R. and Worthington, T. 1974: *New Agricultural Landscapes.* Cheltenham: Countryside Commission.

——1984: *Agricultural landscapes: A second look.* Cheltenham: Countryside Commission.

White, A. and Silverwood, M. 1983: Farming on the urban fringe. *Town and Country Planning* **52**, 113–14 and 150–1.

White, D. 1978: Have second homes gone into hibernation? *New Society*, 10 August, 286–8.

White, P.E. 1974: *The social impact of tourism on host communities: A study of language change in Switzerland.* Oxford: School of Geography.

White, P.E. and Woods, R.I. (eds.) 1980: *The geographical impact of migration.* London: Longmans.

Wibberley, G.P. 1959: *Agriculture and urban growth.* London: Michael Joseph.

——1972: Conflicts in the countryside. *Town and Country Planning* **40**, 259–64.

——1976: Rural resource development in Britain and environmental concern. *Journal of Agricultural Economics* **27**, 1–16.

——1981: Strong agricultures but weak rural economics: The undue emphasis on agriculture in European regional development. *European Review of Agricultural Economics* **8**, 155–70.

——1982: Public pressures on farming: The conflict between national agricultural and conservation policies. *Farm Management* **4**, 373–9.

——1984: *Countryside planning: A personal evaluation.* Ashford: Wye College.

Wilkes, K. 1983: The 1980 Swiss Census. *Geography* **68**, 61–4.

Willatts, E.C. and Newson, M.G.C. 1953: The geographical pattern of population changes in England and Wales, 1921–1951. *Geographical Journal* 119, 431–54.

Williams, G. 1984a: Rural advance factories: A programme in search of a policy. *The Planner* **70**, 11–13.

——1984b: Development agencies and the promotion of rural community development. *Countryside Planning Yearbook* **5**, 62–86.

Williams, H. 1976: Three types of Welsh rural community. *Sociologia Ruralis* **16**, 279–90.

Williams, R.H. (ed.) 1984: *Planning in Europe.* London: George Allen & Unwin.

Willis, K. 1982: Green Belts: An economic appraisal of a physical planning policy. *Planning Outlook* **25**, 62–9.

Winegarten, A. and Acland-Hood, M. 1978: British agriculture and the 1947 Act.

Journal of the Royal Agricultural Society of England **139**, 74-82.

Winsberg, M.D. 1980: Concentration and specialisation in United States agriculture 1939-78. *Economic Geography* **56**, 183-9.

Wolpert, J. 1964: The decision-making process in a spatial context. *AAAG* **54**, 537-58.

Wood, A.P. and Smith, W. (eds.) 1982: *Review and directory of rural geography in the Commonwealth*. Toronto: Department of Geography, University of Toronto.

Woodruffe, B.J. 1976: *Rural settlement policies and plans*. London: Oxford University Press.

Woods, A. 1984: *Upland landscapes change: A review of statistics*. Cheltenham: Countryside Commission.

Woods, K.S. 1968: Small scale industries in the rural and regional economy today. *Town Planning Review* **39**, 251-61.

Woods, R. 1979: *Population analysis in geography*. London: Longman.

——1982: *Theoretical population geography*. London: Longman.

Woollett, S. 1981: *Alternative rural services*. London: Bedford Square Press.

Wormell, P. 1978: *Anatomy of agriculture*. London: Harrap.

Worthington, T. 1982: Agricultural land quality and planning. *Journal of Planning and Environment Law*, September, 561-5.

Zetter, J.A. 1974: The application of potential surface analysis to rural planning. *The Planner* **60**, 544-9.

Index

Authors not appearing in the text but referred to as '*et al.*' have the lead author in brackets at the end of their entry. *indicates the relevant page(s) if this work is cited more than once.

ACC 105
accessibility 93, 97–103, 105, 112, 118, 159
Acland-Hood, M. 142
Action Plans 148
activity rates 91
Adams, J.G.L. 89
Adams, W.M. 167
ADAS 9, 19, 163* (Countryside Commission)
ADC 148
advance factories 147–8, 149, 150
Advisory Council for Agriculture and Horticulture 163
age–sex structure 71, 90–1, 106, 132, 147
Agricultural Adjustment Act 144
agricultural censuses 9–11, 122
Agricultural Geography Study Group 5, 6
agricultural grants and subsidies 12, 14, 132, 142, 143, 167–8
Agricultural Land Service 133
agricultural policy 140–6
agriculture 5, 6, 7, 9–37, 45, 89, 91, 92, 93, 112, 114, 123, 124, 125, 127, 128, 130, 132, 138–9, 140–6, 148, 149, 151, 154, 155, 163–8
Agriculture Act 142
Agriculture and Food Act 144
air photos 11, 122, 123, 124
Aitchison, J.W. 23
Alaska 159
Allaby, M. 132
Allan, J.A. 11, 123
Alps 41, 115, 191, 192
America (*see also* USA) 25, 34, 71, 89, 116, 120, 124–5, 143, 155, 159, 175, 176, 177, 178, 188, 192, 195
Ambrose, P. 81, 83

Amey, L. 36
Anderson, J.R. 27
Anderson, K.E. 19, 20
Anderson, M.A. 56, 57, 59, 96, 122, 123, 124, 134, 155, 156
Anderson, P. 132
Andreae, B. 36
annual price review 142
Anon 144, 145
Appalachia 150, 184
Appleton, J.M. 137
Area of Outstanding Natural Beauty 63, 123, 124, 134, 155, 156, 161
Arensberg, C.M. 79
Argyll 100
Armstrong, A. 148
Armstrong, P. 127
Ashby, A.W. 79
Ashby, P. 115
Ashton, J. 20, 21
Askew, I. 105
Assisted Areas 147
Atlantic coast 113
Aubrey, P. 23
Australia 145
Austria 116, 189

Baker, S. 12
Balchin, W.G.V. 121
Ball, R.M. 97
Banister, D. 98, 99, 101
Barker, D. 12
Barnes, F.A. 13
Barras, R. 152
Barrett, E.C. 11
Barton, P.M. 165
Baxter, M. 118
Beale, C.L. 74

Bealer, R.C. 83, 87, 89
behavioural studies 26–30, 68, 89, 100, 107, 108, 111–15, 117–18, 119, 127–30, 132, 135–6, 137
Belding, R. 34
Belgium 116, 125, 149
Bell, C. 29, 30, 82, 83, 107, 163 (Newby, M.) (all)
Bell, T. 103
Bentham, G.C. 100, 104
Berry, D. 154
Berry, W.G. 71 (Jones, H.R.)
Best, R.H. 84, 121, 122, 124, 155
Bielckus, C.L. 115
Birch, F. 109
Birch, G. 115 (Ashby, P.)
Birmingham Green Belt 154
Blacksell, M. 30, 56, 57, 108, 123, 127, 134, 154, 155
Blake, J. 93
Blakeley, E.J. 83, 93
Blunden, J. 42–3
Boddington, M.A.B. 59, 133
Bollom, C. 115
Bontron, J.C. 87
Borders 113
Bournemouth 74
Bowers, J.K. 168
Bowler, I.R. 5, 9, 12, 13, 14, 15, 17, 20, 25, 36, 37, 141, 144, 146
Bowman, J.C. 142
Boyd, D.A. 19 (Church, B.M.)
Boyer, J.C. 115
Bracey, H.E. 51, 69
Bracken, I. 152
Bradshaw, T.K. 83, 93
Brant, J. 74
Bray, C. 59
Brecon Beacons 114, 123, 130
Briggs, D.J. 136
Bristow, M.R. 152
Britain (British) 9, 10, 11, 17, 21, 22, 25, 31, 34, 39, 42, 43, 44, 45, 53, 71, 75, 79, 82, 83, 87, 88, 94, 95, 122, 142, 147, 149, 151, 161, 178, 184, 187, 189, 195
Brittany 54, 191
Britton, D.K. 25, 31
Broady, M. 150
Brotherton, D.I. 114, 117, 119, 124, 155, 156, 159
Brown, B. 70, 72 (Dean, K.) (both)
Brown, D.L. 71
Brown, J.E. 87
Browning, C. 91
Bruce, A. 122, 130 (Parry, M.L.) (both)

Bruton, M. 152
Bryden, J. 150
BTA 109
Buckley, G.P. 165
Bunce, M. 47
Burdge, R.J. 79, 80
Burnell-Held, R. 138, 151, 152, 168
Bury St Edmunds 57, 59
bus services 97–103
Butler, J.E. 69
Buttel, F.H. 83, 145
Buyhoff, G.J. 134, 135, 136

Caird, J.B. 71 (Jones, H.R.)
California 80, 83, 154
Cambridgeshire 74, 128
Campbell, D. 41
Campbell, R.R. 71
Canada 42, 104, 116, 124, 125, 145, 156, 188–9, 190, 191
Cardiff 56, 100
Carlson, A.A. 136
Carmarthen 92
Carolina 149
car ownership and use 85, 97, 103, 108, 109, 112, 115, 118
Carpenter, E.H. 60
Carroll, M.R. 41
Carruthers, A.O. 74
carrying capacity 157
Carter, C.J. 159
CAS 40
Cathie, J. 129
Ceiber, S. 103 (Bell, T.)
central place theory 48–52, 104–5, 153
Champion, A.G. 70, 74, 76, 124–5, 155
Chapius, R. 87
Chapman, K. 50
Cherry, G. 139
Cheshire, P. 168
Cheviots 133
Chilterns 10, 180
Chisholm, M. 15, 19, 33, 47
Christaller, W. 48, 49
Church, B.M. 19
Churcher, E.H. 34
circular and cumulative causation 70, 90, 91, 145
CLA 163*, 164, 165 (Countryside Commission)
Clark, C. 47, 48
Clark, D. 60, 64, 65, 97, 105
Clark, G. 5, 6, 10, 11, 20, 64
Clawson, M. 40, 44, 117, 151

Clemenson, H. 154
Cloke, P.J. 2, 7, 52, 53, 55, 56, 57, 59, 66, 85, 86, 87, 107, 114, 138, 154, 171, 172
Clout, H.D. 4, 5
cluster analysis 20, 24-5, 29, 51, 61-3, 85, 87, 88, 107
Cmnd 6020 140
Cmnd 7458 142
Cmnd 8804 143
Cmnd 8963 143
Cmnd 9111 150
Coates, V.T. 54
coal 43-4, 152
Cobham, R. 130, 164, 165, 167
Coleman, A. 11, 40, 121, 122, 123, 155
Coleman, R. 112
Colenutt, R.J. 118
Coles, O.B. 101 (Moseley, M.J.)
Collins, N.J. 43
Colorado 149
Commonwealth 6, 7, 144
Commins, P. 91, 145
Commission of the EEC 21, 22, 142, 169
Commission on Energy and the Environment 44
Common Agricultural Policy 142, 143, 144, 145, 148-9
communities 79-83, 106, 115, 145, 148, 149, 169
commuting 62-3, 81, 97
comparative advantage 140
Compton, P.A. 72
conflict resolution 139-40, 151-63
Connell, J. 81, 83
conservation 6, 41, 59, 63, 130, 156-63, 163-8, 169
Conservation Foundation 159
Conservative Party 107
Cooke, P. 150
Coombes, M.G. 93 (Owen, D.W.)
Coppock, J.T. 6, 10, 13, 16, 18, 19, 25, 31, 32-3, 36, 108, 113, 117, 120, 121, 122
Corn Belt 15, 156
Cornwall 70, 91, 196
COSIRA 147, 148, 149
cost-benefit studies 40-1, 45-6, 49-51, 117, 151
Cotswolds 43, 60, 61
Cottam, M.B. 107
Cotton Belt 15
Coughlin, N. 45
council houses 64-5, 81, 82, 154-5
Council of Europe 163

counterurbanization 70-9, 91-96, 104
country parks 5, 114, 157
Countryside Act 114, 165, 167
Countryside Commission 93, 108, 109, 114, 118, 127, 129, 131, 132, 157, 161, 163, 164, 166, 167, 168, 169, 173
Countryside Commission for Scotland 114, 137, 161
Countryside Planning Yearbook 2, 171-2
county councils 57, 99, 148, 151, 152, 154, 169
County Public Transport Plans 99
Cowap, C. 167
Cracknell, B.E. 20, 21, 118
Craig, J. 84, 85, 87, 88
Craig-y-Nos Country Park 114
CRC 91, 139, 142, 168
Cribier, F. 115, 116
Cromley, R.G. 34
Crosbie, A.J. 74 (Carruthers, A.O.)
Cross, D.T. 152
Cruickshank, A. 67, 72, 73
Cullingford, D. 87*, 107 (Openshaw, S.)
Cullingworth, J.B. 151
Cunningham, R.A. 154
Curry, N. 59, 117
Curtis, L.F. 11, 164

Daft, L.M. 105
Dale, I. 39, 40
Darby, J. 91
DAFS 13, 89
DART 116
Dartmoor 43, 123, 130, 133, 162
Dartmoor NPA 162
Davidson, J. 138, 163, 170
Davies, E.T. 114
Davies, M.C. 79
DBRW 148
Dean, C. 104
Dean, K. 70, 72
Deane, G.C. 123
Dearden, P. 135, 136
definition of rural areas 4, 83-9, 97, 172
definition of rural settlements 47-8
De Jong, G.F. 71
demand curve analysis 117
demonstration farms 164, 167
Dench, S. 91
Denham, C. 84, 85
Denman, R. 114
Denmark 116, 125, 152
Dennier, A. 157

Dennis, S.J. 118
density size rule 84
depopulation 68–79, 87, 90, 106, 115, 150
deprivation 57, 59, 61, 64–5, 97, 101, 104, 105–7
DES 39
De Vane, R. 115
developing countries (*see* Third World)
Development Commission 147, 148, 149
development control 56, 64, 151, 152, 154–6
Development Plans 53–60, 84, 122, 151
Devis, T. 67
Devon 54, 56, 65, 91, 92, 93, 154, 155, 156, 157, 177, 194, 196
Dickenson, R.E. 51
Dickinson, G.C. 121, 122
Dillard, J.E. 154
Dillon, J.L. 27 (Anderson, K.E.)
Dinse, J. 139
discriminant analysis 19, 20, 27
district councils 57, 148, 152
DOE 44, 61, 121, 122, 123, 151, 152, 157, 164, 165, 168
Doren, C.S. 120
Dorset, 20, 81, 104, 128, 188, 196
Dower, M. 116
Downing, P. 116
Doyle, C. 40, 140
Doyle, C.J. 142 (Bowman, J.C.)
Drewett, R. 155 (Hall, P.)
Drudy, P.J. 71, 90, 91, 145
Drudy, S.M. 71, 91, 145
Duffield, B.S. 6, 108, 113, 120
Duffy, P.J. 76
Dunn, M. 61, 62, 63
Dyfed 100

Eadie, J. 169 (Maxwell, T.J.)
easements 154, 168
East Anglia 26, 27, 28, 30, 54, 72, 82, 191
East Midlands 43, 72
East Sussex 56
ecological sustainability 169
ecology 156–63, 169–70
economies of scale 53–60, 65
Edinburgh 111, 113
education 98, 101, 104, 105, 109
Edwards, A. 31
Edwards, C.J.W. 23, 24
Edwards, J.A. 49, 67
Edwards, S.L. 118

EEC 10, 34, 93, 124, 132, 140, 142, 143, 144, 145, 148, 152, 176
Elson, M.J. 154
Emerson, D. 104
employment 5, 25–6, 54, 67–89, 89–96, 105, 106, 139, 145–50
Employment, Department of 89
England (English) 12, 38, 43, 45, 49, 60, 69, 74, 147, 148, 178, 192
England and Wales (English and Welsh) 14, 16, 18, 19, 21, 22, 23, 32–3, 53, 55, 62, 76, 77, 85, 86, 114, 122, 123, 133, 157, 165, 175, 176, 177, 178, 189, 195
Engledow, F. 36
Environment, Department of (see DOE)
environmentalism 4
environmental determinism 1
Environmental Impact Assessment 152
Errington, A. 26
Es, V.J.C. 87
Essex 60
Estall, R. 149, 150
Europe 9, 10, 25, 72, 79, 93, 115, 116, 123, 124, 140, 152, 163, 175, 187, 189, 198
European Communities 142, 189
Evans, J.A. 19 (Church, B.M.)
Everett, R.D. 117
Ewing, G. 118
Exmoor 123, 130, 133, 164, 182, 194
Expenditure Committee 168
experimental management schemes 169

factor analysis 89, 134–5
farm enterprises 12–26, 31, 32–3
farm rationalization 25
farm tourism 114–15
farm workers 25–6, 65, 82–3, 89, 90, 107, 142
farmers 26–30, 82–3, 89, 90, 107, 127, 132, 139, 142, 163, 167, 168
Farrington, J.H. 100, 101
Feist, M. 164
Fens 25
Ferguson, M.J. 112
Fielding, A. 72
Fife 113
Finland 116, 169
Fisher, D.W. 120
Fisher, J.S. 96
Fitton, M. 112
Flaherty, M. 134
Fliegel, F.C. 71

Flinn, W.L. 83
Florida 154
Foeken, D. 71
Food and Agricultural Organization 10
Ford, P.J. 71 (Jones, H.R.)
Fordham, R. 122
forestry 5, 38–42, 91, 92, 123–4, 125,
    130, 131, 132, 133, 148, 169
Forestry Commission 38, 39, 40, 41, 42,
    123, 163* (Countryside Commission)
Forster, R.R. 163
Fothergill, S. 92, 93
Fotheringham, A.S. 27
Found, W.C. 34, 36*
France 25, 87, 115, 116, 123, 125, 149
France, J. 136
Franklin, S.H. 79
Fraser, A. 41
Frost, M. 92, 93, 94, 95
Frosztega, J. 85
Fuguitt, G.V. 47, 69, 71, 103, 104, 149
FWAGS 163

Gasson, J. 25, 27, 28, 29, 30, 45, 65
Gauthier, L. 136 (Buyhoff, J.G.)
Gayler, H. 156
Georgia 11, 149
Germany 41, 48, 49, 116, 123, 125, 179
Gibbs, R.S. 112
Gilg, A.W. 2, 10, 21, 30, 39, 44, 52, 54,
    56, 57, 60, 67, 91, 92, 93, 114, 127,
    133, 134, 135, 136, 138, 139, 140, 142,
    145, 147, 154, 155, 157, 171
Gilder, I.M. 57, 58, 59
Gillard, A. 87 (Openshaw, S.)
Gillespie, A.E. 93 (Owen, D.W.)
Gillon, S. 64
Glasgow 81, 107, 113
Glasgow, N. 71 (Fliegel, F.C.)
Gloucestershire 103
Glyn-Jones, A. 51, 93
Goldschmidt, W. 25
Goodall, B. 41
Gordon, D.S. 11
Goyder, J. 139
Gracey, H. 155 (Hall, P.)
Grafton, D. 57, 69
gravity models 118
Great Lakes 149
Great Plains 34
Green, B. 156, 170
Green, P.M. 104
green belts 112, 154
Gregor, H.F. 31

Gregory, D.G. 154
Gregory, R. 43
Grigg, D.B. 25, 30–1, 36, 89
Grossmann, D. 52
growth centres 53–60, 145, 148, 150
guaranteed prices 142, 143
Gudgin, G. 92, 93
Guldmann, J.M. 134

Haffey, D. 112
Hägerstrand, T. 12, 27, 36
Haining, R. 52
Hall, P. 121, 155
Hamersley, C. 164
Halsall, D.A. 97
Hamilton, P. 167
hamlets 47–60
Hannan, D.F. 69, 91
Hanrahan, P. 57, 138
Hardaker, J.B. 27 (Anderson, K.E.)
Harding, P.T. 124
Harkness, C.E. 122*, 124, 130* (Parry,
    M.L.)
Harman, R. 101*, 103, 104 (Moseley,
    M.J.)
Harris, R. 123
Harrison, C. 112
Hart, J.F. 10–11, 121, 155, 158
Hart, P. 163
Hart, P.W.E. 30, 34
Harvey, D.W. 3, 4, 6, 12, 26
Hassinger, E. 69
Haslett, M. 115 (Ashby, P.)
Hathout, S.A. 123
Haynes, R.M. 100, 104
Hawaii 154
HC 98, 168
health and welfare 7, 100, 101, 104, 105
Healey, P. 152, 155
Healy, R.G. 154
Hedger, M. 93
Herefordshire 82, 98, 128, 188, 189
Heritage Coasts 161
HIDB 147, 148, 149, 150
Higgs, A.J. 156 (Margules, C.)
Highlands and Islands 74, 101, 113, 148,
    149, 175, 178, 188, 189, 192
High Weald 175
Hilgendorf, L. 65
Hill, B. 25
Hirsch, G.P. 25
HL Select Committee:
    On Science and Technology 40
    On Sport and Leisure 108

On the European Communities 143, 148, 149
HMSO 9
Hockin, R. 164
Hodge, M. 70, 90, 92
holidays 109
Holland 44, 116*, 123*, 125*, 152, 168, 178*, 197* (* *under* Netherlands)
Holtam, B. 42
Hop Marketing Board 12
Hornby, R.W. 30
Hoskins, W.G. 127
household deprivation 105
House of Commons (*see* HC)
House of Lords (*see* HL)
housewives 102, 103
housing 5, 56, 60–6, 81, 105, 106, 115–7
Housing Act 64
housing associations 65
Housing and Local Government, Ministry of 151–2
Hudson, J.C. 52
Hudson, R. 122
Huggett, R. 36
Hughes, J.T. 51
Hull, D. 134
Hultmann, S. 136 (Buyhoff, J.G.)
humanism 137
Humphrey, C.R. 71
Huntingdonshire 128

IBG 5
Ilbery, B.W. 12, 20, 26, 27, 28, 29, 30, 36
Illinois 89
impact studies 111–15
index of rurality 85–7
informal recreation 112
infrastructure 7, 47–60, 148
innovation diffusion 12, 27, 36, 37, 164
integrated rural development 140, 149, 150, 168–70
Intervention Board for Agricultrure 143
intervention system 142–3
Iowa 52, 194
Ireland 71, 76, 116, 123, 125, 175, 183, 184
Irving, B. 65
Isbell, J. 11 (Coleman, A.)
island-biogeographic theory 157
Italy 116, 123, 125
ITE 130
IUCN 169

Jackson, R.H. 123, 124, 126, 127, 154
Jacques, D.L. 136
Japan 44, 145
Johansen, H.E. 47, 71, 103, 104
Johnson, D.M. 71
Johnson, H.B. 137
Johnston, R.J. 1, 69, 70
Jones, A. 65
Jones, A.R. 84 (Best, R.H.)
Jones, D.W. 34
Jones, H.R. 67, 70, 71
Jones, P. 104
Joseph, A.E. 104
Journal of Rural Studies 2

Kariel, H.G. 115
Kariel, P.E. 115
Keeble, D. 93
Kennett, S. 67
Kent 25, 59, 187
key settlements 53–60, 63, 66, 102, 145, 154–5
Killeen, K. 134
Kimball, S.T. 79
Knowles, D.J. 10 (Clark, G.)
Knowles, R. 97
Knox, P.L. 107
Knulesky, W.P. 83, 87, 89 (Bealer, R.C.) (all)
Koch, N.E. 136 (Buyhoff, J.G.)
Kostrowicki, J. 15
Kozlowski, J. 51
Kreimer, A. 135
Kusler, J.A. 168

labour markets 90
Labour Party 140
Lake District 64, 133, 169, 179, 180, 181
Lance, A.N. 132
land classification maps 133–4
Land Decade Educational Council 121
land evaluation 132–134
land lost from agriculture 44–5, 125, 126, 151, 154, 155, 158
land management 140, 163–70
land use 5, 44–6, 121–37, 151–63
land use maps and surveys 11–12, 84, 121, 122, 155
landscape 41, 64, 115, 121–37, 155, 156–63, 163–8
landscape evaluation 134–7
Lang, J.T. 159
Lapping, M.B. 154

Lassey, W.R. 139, 151
Laurie, I.C. 134 (Robinson, D.G.)
Lavery, P. 118, 120
Law, C.M. 71
Lawton, R. 97
Leat, D. 139
Lefaver, S. 151
Leicester 118
Leonard, P. 124, 130, 164
Less Favoured Areas 114–15, 147, 148–9
Lewis, G.J. 68, 70, 76, 78, 79, 82, 97
Lewis, J. 120 (Doren, C.S.) (Fisher, D.W.)
Liberal Party 140
Liddle, M.J. 112
linear logistic model 118
Linton, D.L. 134
de Lisle, D.G. 34
Lloyd, R. 163, 164
Local Government Act 99
Local Plans 152
location theory 33–6, 47, 91–2, 105
logical positivism 2–3, 72, 108*, 120 (* *under* scientific route)
London 25, 44, 45, 56, 112, 154, 186, 192
Lonsdale, R. 52, 91
Lösch, A. 48, 49
Lovejoy, D. 40
Low, N. 44
Lowe, P. 139
Lowenthal, D. 137
lowland landscapes 127–30
Luloff, A.E. 4, 89
Luxembourg 116, 125, 149
Lyons, E. 136

McCallum, J.D. 89
McCoy, E.D. 157
McCuaig, J. 124
MacEwen, M. 159, 167
McLaughlin, B.P. 59, 107, 112
MacLeary, A.R. 139
McNab, A. 168
McNamara, P. 155
Mabogunje, A. 68
MAFF 9, 14, 16, 21, 22, 23, 25, 45, 122, 133, 163–8
Malaysia 30
Mallet, A. 115
management plans and agreements 157–9, 163–8
Mandelker, D.R. 154
Manners, G. 44

Manning, E. 124
Maos, J. 105
marginal land (*see* uplands)
Margules, C. 157
market towns 47–60, 82, 91, 147–8
Marshall, I.B. 42
Martin and Voorhees 56, 154
Martin, W.H. 117
Mason, S. 117
Mather, A.S. 38, 39
Matthieu, N. 87
Maund, D.J. 76, 78, 82
Maunder, A.H. 25
Maxwell, T.J. 169
Meinig, D.W. 137
Merioneth 92
Merrett, S. 60
Meyer, I. 36
Mid Wales Development 148, 149
middle west 15
migration 67–89, 147
Miles, J.C. 118
Miles, R. 41
Milk Marketing Board 12–13
Miller, A.K. 157
Miller, M.K. 4, 89
Milton of Campsie 81, 82
Minay, C. 148
mining 42–4
Minister of Land and Natural Resources 108
Ministry of Rural Affairs 139–40
Minnesota 69
Mitchell, G.D. 80
Mitchelson, R.L. 96
mobility deprivation 105
models and theory 26–7, 33–6, 48–50, 52–60, 68, 72, 78, 84, 100, 105, 117–19, 120
Mogey, J. 79
MOH 61
Montgomery 92
moorland (*see* uplands)
Moran, W. 34
Morgan, W.B. 31
Moseley, M.J. 54, 76, 79, 91, 100, 101, 102, 103, 105, 106, 107
Moss, G. 47, 81, 82
Motorways: M4 Corridor 83; M25 154
Muir, G. 137
Muller, P.O. 34
multiple discriminant analysis 89
multiple land use 41, 43, 44–6, 140, 157, 167, 168–70
multivariate statistics 20, 24–5, 29, 51,

61–3, 85, 87, 89, 107, 117–18, 134–5, 136

Munton, R.J.C. 19, 31, 44, 45, 112, 154

National Bus Company 99
National Consumer Council 98
National Parks 5, 63–4, 93, 109, 114, 123, 130, 140, 155, 157–9
National Park Plans 157–9
National Scenic Areas 161
National Travel Survey 98
National Union of Agricultural Workers 65
nature reserves 112, 117, 157, 161
NCC 132, 157, 160, 163, 163*, 170 (Countryside Commission)
NCVO 91, 150
Neate, S. 105
need studies 100–1
Nelson, J.G. 163
Netherlands 116, 123, 125, 178, 197
Netting, R.M. 169
de Neufville, J.I. 151
new agricultural landscapes 127–30, 163–7
New Ash Green 59
Newbury, P.A.R. 36
Newby, H. 26, 29, 30, 82, 83, 107, 163
New England 150
New Mexico 149
Newson, M.G.C. 69
Newtown 70, 148
New Towns 83
new villages 59–60
New Zealand 145
NFU 163,** 164* (CLA)** (Countryside Commission)*
Niagara Fruit Belt 156
Nicholls, D.C. 41
Nicholson, B. 71
Norfolk 49–51, 74, 90, 103, 112, 183, 186, 190, 192
Norfolk County Council 49
Norfolk County Planning Department 4, 103, 104
Norris, J.M. 19
North, J. 44
North Sea Coast 113
Northumberland 92, 130
North York Moors 123, 130, 133
Norway 70–1, 116
Nottinghamshire 80
Nutley, S.D. 100, 101
Nyberg, K.L. 7 (Picou, J.S.)

OECD 140, 144, 151
Olwig, K. 137
Ontario 27, 185
OPCS 67, 76, 84, 88
Openshaw, S. 87, 107
opportunity deprivation 105
Ordnance Survey 122, 124
O'Riordan, T. 169, 170
Orwin, C.S. 139
Owen, D.W. 93 (Keeble, D.L.)
Owens, P.L. 108, 118, 120
Oxfordshire 29, 60, 61, 74, 188
Ozarks 149

Pacione, M. 2, 6, 8, 51, 61, 76, 81, 82, 83, 106, 116, 143
Pahl, R.E. 61, 79, 80, 81, 82, 83
Pallone, R. 72
Palmer, C.J. 89
Park, C.C. 107, 114
Parry, M.L. 122, 123, 124–5, 130
part time farming 23–5
party and local politics 82–3, 107, 139–40, 163
Passenger Vehicles (Experimental Areas) Act 99
Patmore, J.A. 113, 120
Peak District 43, 133, 153, 157, 169, 175
Peak Park Joint Planning Board 153, 157, 159, 169
Peake, H. 69
peasantry 79
Peat, Marwick, Mitchell & Co. 100, 101, 103
Pennines 133
Penning-Rowsell, E.C. 134, 135, 169
Penoyre, J. 47
Perry, R. 70, 72 (Dean, K.) (both)
Peterken, G.F. 41, 124
Peters, G.H. 121
Peterson, G. 156
Phillips, D. 2, 8, 56, 60, 61, 64, 65, 66, 97
Phillips, Derek 173
Phillips, H.L. 10 (Clark, G.)
Picou, J.S. 7
Pierce, J.T. 155
Pigram, J. 110–11, 159
Pitzl, G.R. 137
planning 5, 6, 7, 43–4, 45, 52–60, 84, 130, 134, 138–70
Platt, R.H. 156
Plaut, T. 154
Pocock, D.C.D. 12, 137
Poland 25

population 5, 6, 7, 53–60, 67–89
population census 61–3, 67, 84, 85, 87, 107
population change 67–79, 90
population density 84–5, 89, 155
population potential 85
Porchester, Lord 164
Portugal 116
post-industrial society 83
post offices 104
potential surface analysis 157–68
Potter, C. 168
Preece, R.A. 134, 135
pressure groups 83
Price, C. 39, 40, 117, 136
Priddle, G. 120
primary employment 25–6, 83, 90, 145–50
principal components analysis 19, 20, 29, 51, 85, 117, 135
Probert, G. 164
Propst, D.B. 135
prospect–refuge theory 137
public inquiries 43–4, 152
public transport 97–103

quotas 142, 144

Rackham, O. 41
Radford, E. 81, 83
Rafe, R.W. 156 (Margules, C.)
railways 97, 99
Ramblers Association 41
Randolph, W.G. 72, 75, 76, 77
Ravenstein, E.G. 68, 69, 70, 74, 172
Rawson, M. 61, 62, 63 (Dunn, M.) (all)
Rayner, A.J. 13
RCCs 148
recreation 5, 6, 7, 41, 105, 108–20, 156–63, 169
Reeds, L.G. 27
Rees, G. 67
Rees, T.L. 67
Regional Fund 149
regional multiplier 109, 115, 145
Regional Planning Commissions 149–50
regional policies 145–150
regression analysis 84, 101, 117, 118, 134, 135, 136
Relph, E. 137
remote sensing 11, 123
Rent (Agriculture) Act 65–6
repopulation 59, 68–79, 87, 96

resale of council houses 64–5
restraint policies 55
resource development 140–50
retirement 71, 79, 81, 103
Rettig, S. 45
RICS 165 (CLA)
Rhind, D. 67, 122
Richmond, P. 65
Rieger, J.H. 91
Ringmer 81
Robbins, C.J. 142 (Bowman, J.C.)
Robert, S. 72, 75, 76, 77
Roberts, B.K. 47
Robertson, I.M.L. 83, 84, 100
Robinson, D.G. 134
Robinson, M.E. 89 (Palmer, C.J.)
Rogers, A.W. 31, 60, 61***, 62, 63, 64, 84*, 115**, 134 (Best, R.H.)* (Bielckus, C.L.)**, (Dunn, M.)***
Rogers, E.M. 79, 80
Roome, N.J. 117
Rose, P. 29, 30, 82, 83, 107, 163 (Newby, H.) (all)
Rosenberg, J.S. 154
Rosenblood, L. 135
Rowley, G. 104
Rowley, T. 47
Roxburgh 92
Royal Geographical Society 11
Runte, A. 159
Rural Development Areas 147
Rural Geography Study Group 5, 6
Rural Resettlement Group 59
Rural Sociology 7
rural–urban continuum 79–80
Rushton, G. 103 (Bell, T.)
Russia 30

Sadler, J.I. 19 (Church, B.M.)
Sandford Report 157, 159
Sargent, F.O. 154
Saunders, P. 29, 30, 82, 83, 107, 163 (Newby, H.) (all)
Saville, J. 69
Schaefer, F.K. 2
Scorgie, H.R.A. 112
Scotland (Scottish) 13, 15, 21, 38, 41, 51, 69, 74, 83, 89, 101, 107, 111, 113, 123, 130, 147, 148, 149, 168, 178, 182, 188, 189, 197
SDD 51
second homes 5, 79, 106, 115–17, 154
secondary employment 83, 89–93, 145–50

Secretary of State for the Environment 108
Section 52 Agreements 64
Selman, P. 169
semantic difference 89, 135
services 47–60, 92–3, 100, 103–5, 116, 147, 153
Settle, J.G. 118
settlement 5, 6, 7, 47–60, 69, 74, 84, 90, 103
settlement policies 52–60
settlement rationalization 49–52, 53–60, 69, 104–5
Sewell, W.R.D. 6
Shaw, D. 70, 72 (Dean, K.) (both)
Shaw, D.P. 53, 55
Shaw, J.M. 49
Shaw, M.G. 121, 122
Shaw, M.J. 105
Shoard, M. 130, 167
Shropshire 49, 51
Shucksmith, D.M. 64, 115, 116, 117
Shuttleworth, S. 135
Sibbald, A.R. 169 (Maxwell, T.J.)
Silicon Valley 83
Silverwood, M. 45
Simmons, I.G. 37, 120
Simpson, B. 51
Simpson, E.S. 13
Sinclair, G. 11*, 132 (Coleman, A.)
Sinden, J.A. 157
Singleton, P. 112
Sites of Special Scientific Interest 165–7
site surveys 111–15
Slee, R.W. 145, 157
Sluman, B. 169
Smart, G. 103
Smit, B. 104, 134
Smith, N. 118
Smith, S.L.J. 108, 110, 120
Smith, W. 6, 7
Snowdonia 130, 133, 169, 181
social class 61, 71, 72, 80, 81, 87, 89, 109, 136
Social Fund 149
social structure 5, 64, 71, 79–83, 105, 118
Sofranko, A.J. 71 (Fliegel, F.C.)
Somerset 23, 24, 69, 128, 129
South West 70, 72, 91, 109, 114, 116, 118, 130, 147
South West Economic Planning Council 109, 116
South Woodham Ferrers 59–60
Spain 116
Spence, N. 67, 92, 93, 94, 95

Spencer, M.B. 101 (Moseley, M.J.)
Spooner, D.J. 42, 43, 44, 91
Stamp, L.D. 11, 121, 122
standard man-days 21–2
Standing Conference of RCCs 103, 104
Stanley, P.A. 100, 101
Stewart, A.J.A. 132
Strathclyde 51, 82, 195
Strauss, E. 34
Strong, A.L. 154, 168
Structure Plans 53–60, 151, 152
Sturrock, F.G. 129
Suffolk 57, 74
sunrise industry 70–6, 91–2, 93
Surrey 74, 81, 180
Sussex 56, 81, 155, 156, 175
Sutherland 92
Sweden 116, 125, 130
Switzerland 60, 116, 152, 169, 192, 198
Symons, L. 30, 31

Tandy, C. 41
Tarrant, J.R. 26, 31, 35, 36, 144
Tavener, L.E. 20
Taylor, C. 104
tertiary employment 83, 89–93, 145–50
Third World 6, 144
Thissen, F. 115
Thomas, R. 155 (Hall, P.)
Thomas, R.W. 89 (Palmer, C.J.)
Thompson, C. 93 (Keeble, D.L.)
Thomson, K.J. 45
Thorneycroft, W. 70, 72 (Dean, K.) (both)
Thorns, D.C. 80, 81, 83, 139
thresholds 48–52, 103, 104
Thurgood, G. 66
tied housing 64–66, 82, 87
time–space geography 100, 102
Tonnies, F. 79
tourism 6, 7, 91, 93, 108–20, 149
Towler, R.W. 41
Town and Country Planning Acts 64, 151, 154, 164
Tracy, M.A. 140
Traill, A.L. 134 (Robinson, D.G.)
transport 5, 6, 84, 85, 91–2, 93, 97–103
Transport Acts 99
Transport, Department of 99
Transport Policies and Programmes 99
Tranter, R. 140
Treasury, H.M. 40, 41, 92
Treaty of Rome 142
Trimble, S.W. 11

TRRU 93, 109, 111
Turnock, D. 69
Turton, B.J. 97
type of farm maps 13–20
typologies 29–30, 80–3, 85–7

Ulster Countryside Committee 161
United Kingdom 6, 10, 22, 38, 39, 43, 60,
    67, 79, 83, 87, 93, 96, 103, 116, 122,
    123, 124, 125, 138, 139, 140, 143, 144,
    145, 151, 154, 155, 156, 157, 168, 169,
    177
United Nations 10
Unwin, K. 97
Uplands (moorlands and marginal areas)
    112, 114–15, 122–3, 130–3, 145,
    148–9, 163, 164
urban fringe 44–6, 56, 112, 154, 155, 164
USA 10–11, 34, 40, 44, 45, 47, 60, 67, 69,
    71, 72, 73, 74, 91, 93, 96, 103, 104,
    105, 116, 123, 124, 125, 126, 138, 139,
    140, 142, 143, 144, 145, 149, 151, 154,
    155, 156, 158, 159, 168, 178
US Department of Agriculture 124
Usher, M.B. 157
Utah 149

Vale of Belvoir 43, 183, 185
Vermont 154
villages 47–60, 65, 69, 71, 74, 79, 80, 81,
    82, 103, 104, 154–5
Vince, S.W.E. 69
Vining, D. 72
visitor surveys 109, 111–15
Visser, D.W. 138, 151, 152, 168
Visser, S. 33, 34
Von Thünen 33, 34, 35, 37, 172

Wager, J. 134 (Robinson, D.G.)
Wagstaff, H.R. 25
Wales (Welsh) 23–4, 38, 40, 41, 70, 72,
    74, 82, 93, 100, 114, 116, 123, 130,
    133, 148, 149, 150, 175, 176, 177, 182,
    187, 189, 193, 195, 198
Walker, A. 105
Walker, A.J. 164
Wardwell, J.M. 71
Warford, J. 49, 51, 105
Warner, W.K. 83
Warnes, A.M. 71
Warwickshire 56, 128, 129, 134, 154, 188
Watkins, C. 38, 40
Waugh, T.C. 74 (Carruthers, A.O.)

Weaver, J.C. 15, 16, 19
Webber, R.J. 87, 88
Weils, R.H. 7 (Picou, J.S.)
Wellman, J.D. 136 (Buyhoff, J.G.)
Welsh Development Agency 150
Westmacott, R. 127, 128, 129
West Midlands 12, 27, 28, 188
Western Isles 89, 107, 149, 168
Wheeler, P.T. 38, 40
Whitby, M. 70, 90, 92
White, A. 45
White, D. 117
White, P.E. 70, 115
Wibberley, G.P. 4, 115*, 138, 139, 145,
    151, 168, 170 (Bielckus)*
Wigtown 92
wildlife 112, 117, 132, 156–63, 163–8,
    169
Wildlife and Countryside Act 164, 165,
    168
Wilkes, K. 60
Willatts, E.C. 69
Williams, A. 2, 8, 56, 60, 61, 64, 65, 66,
    97
Williams, G. 147, 148, 149, 150
Williams, H. 82
Williams, R.H. 152
Willis, K. 45
Willits, F.K. 83, 87, 89 (Bealer, R.C.)
    (all)
Wiltshire 74, 103
Winegarten, A. 142
Winsberg, M.D. 34
Wisconsin 69, 83
Wolpert, J. 26
Wood, A.P. 6, 7
Woodruffe, B.J. 53, 87
Woods, A. 130, 131, 133
Woods, K.S. 91
Woods, R.I. 67, 68, 70
Woolcock, J. 167
Woollett, S. 103, 105
Worcestershire 81
World Conservation Strategy 169
Wormell, P. 36
Worthington, T. 127, 128, 129, 133
Wright, S. 103

Yalden, D.W. 132
young people 90–1, 104
Yorkshire 43, 128
Yorkshire Dales 113

Zetter, J.A. 157
zoning 152, 154, 156, 159